Growing with Mathematics

Student Book

Volume 1

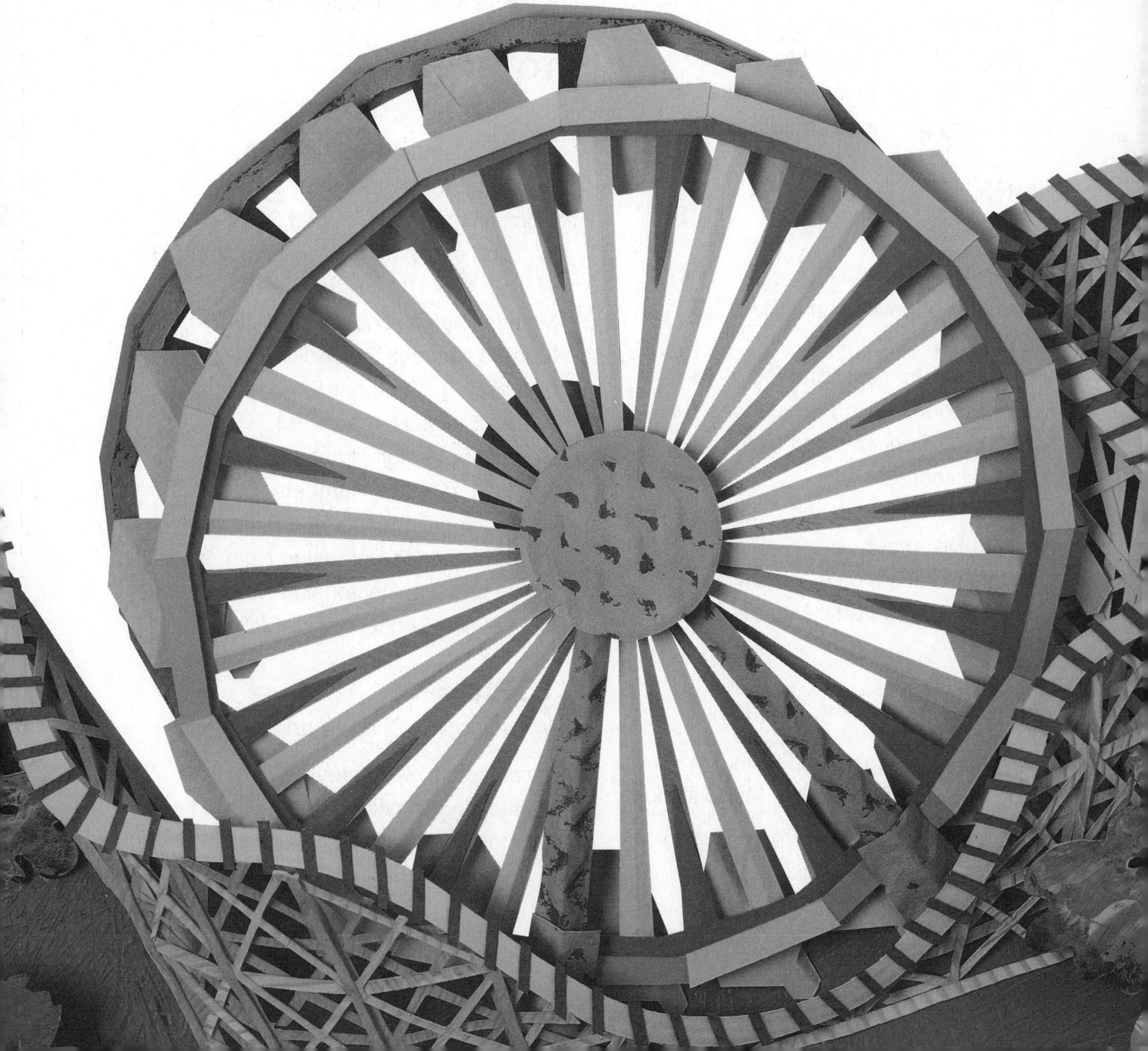

Contents

TOPIC 1
Exploring Mathematics

Maintaining Concepts and Skills 1
Counting On and Counting Back 2
Maintaining Concepts and Skills 3
Counting On and Counting Back 4
Maintaining Concepts and Skills 5
Adding 1-digit and 2-digit Numbers 6
Adding Amounts 7
Working with Parallel Lines 8
Adding 1-digit and 2-digit Numbers 9
Reading and Creating Bar Graphs 10
Maintaining Concepts and Skills 11

TOPIC 2
Investigating Length

Testing Triangles 12
Maintaining Concepts and Skills 13
Relating Distances 14
Maintaining Concepts and Skills 15
Reading and Creating Bar Graphs 16
Maintaining Concepts and Skills 17
Using Millimeters 18
Maintaining Concepts and Skills 19
Relating Distances 20
Reading Numbers on a Number Line 21
Reading Numbers on a Number Line 22

TOPIC 3
Using Addition and Subtraction

Maintaining Concepts and Skills 23
Using Addition, Subtraction, Multiplication, and Division 24
Investigating Distances 25
Comparing Coastlines 26
Adding and Subtracting 3-digit Numbers 27
Addition and Subtraction Problems 28
Maintaining Concepts and Skills 29
Subtracting 3-digit Numbers 30
Maintaining Concepts and Skills 31
Subtracting 3-digit Numbers 32
Maintaining Concepts and Skills 33
Addition and Subtraction Problems 34

TOPIC 4
Investigating 5-digit Numbers

Using a Number Expander 35
Creating 4-digit Numbers 36
Maintaining Concepts and Skills 37
Reading and Writing 4-digit Numbers 38
Writing About Thousands 39
Maintaining Concepts and Skills 40
Using a Number Expander 41
Completing a Number Expander 42
Writing 5-digit Numbers 43
Comparing 5-digit Numbers 44
Comparing 5-digit Numbers 45
Maintaining Concepts and Skills 46
Increasing and Decreasing Numbers 47
Increasing and Decreasing Numbers 48
Maintaining Concepts and Skills 49

TOPIC 5
Investigating Shapes and Angles

Investigating Symmetry 50
Maintaining Concepts and Skills 51
Maintaining Concepts and Skills 52

Home Link page numbers are shown in gray.

Using a Map 53
Maintaining Concepts and Skills 54
Creating and Investigating Triangles 55
Investigating and Creating Angles 56
Identifying Shapes 57
Identifying Angles 58
Maintaining Concepts and Skills 59

TOPIC 6

Finding Fractions

Using Fractions 60
Maintaining Concepts and Skills 61
Finding a Fraction of an Amount 62
Finding a Fraction of an Amount 63
Solving Fraction Problems 64
Maintaining Concepts and Skills 65
Creating Fractional Parts 66
Maintaining Concepts and Skills 67
Finding Equivalent Fractions 68
Finding Equivalent Fractions 69
Creating Equivalent Fractions 70
Maintaining Concepts and Skills 71

TOPIC 7

Multiplying by 1-digit Numbers

Creating Multiplication Arrays 72
Maintaining Concepts and Skills 73
Practicing Multiplication Number Facts 74
Practicing Multiplication Number Facts 75
Identifying Numbers 76
Maintaining Concepts and Skills 77
Writing and Identifying Factors and Multiples 78
Using Number Facts to Multiply Tens 79
Maintaining Concepts and Skills 80
Using Number Facts to Multiply Tens 81
Maintaining Concepts and Skills 82
Using Arrays to Find Products 83
Multiplying Costs 84
Multiplying Tens 85
Encyclopedia Sale 86
Finding Multiplication Patterns 87
Multiplying 2-digit Numbers 88

TOPIC 8

Working with Time

Using a Calendar 89
Maintaining Concepts and Skills 90
Calculating Time 91
Maintaining Concepts and Skills 92
Creating a Schedule 93
Using Multiplication 94
Relating Units of Time 95
Maintaining Concepts and Skills 96
Maintaining Concepts and Skills 97
Using a Timetable 98
Calculating Time Differences 99

TOPIC 9

Exploring Shapes and Perimeters

Prism Patterns 100
Maintaining Concepts and Skills 101
Finding Perimeter 102
Maintaining Concepts and Skills 103
Estimating and Finding Perimeter 104
Maintaining Concepts and Skills 105
Finding Perimeter 106
Calculating Perimeter 107

Home Link page numbers are shown in **gray**.

Calculating the Perimeter of Polygons **108**
Finding and Comparing Perimeter **109**
Maintaining Concepts and Skills **110**

TOPIC 10

Working with Fractions

Using Fractions **111**
Maintaining Concepts and Skills **112**
Adding and Subtracting Fractions **113**
Adding and Subtracting Fractions 114
Solving Fraction Problems **115**
Maintaining Concepts and Skills **116**
Fractions on a Number Line **117**
Solving Fraction Problems **118**
Using Number Lines Greater than 1 **119**
Maintaining Concepts and Skills **120**
Adding and Subtracting Fractions **121**
Maintaining Concepts and Skills **122**
Adding and Subtracting Fractions 123
Reading and Writing Hundredths **124**
Equivalent Common Fractions and Decimal Fractions **125**
Decimals on a Number Line **126**

TOPIC 11

Linking Multiplication and Division

Problems to Share **127**
Sharing Problems **128**
Relating Multiplication and Division **129**
Relating Multiplication and Division 130
Multiplication Machines **131**
Maintaining Concepts and Skills **132**
Multiplication Machines 133
Super Shirt Factory Sale **134**
Exploring Division Patterns **135**
Maintaining Concepts and Skills **136**
Grab the Cubes **137**
Maintaining Concepts and Skills **138**
Exploring Division Patterns 139
Maintaining Concepts and Skills **140**
Solving Division Problems **141**
Solving Division Problems 142

TOPIC 12

Investigating Large Numbers

Reading Numbers on a Number Line 143
Maintaining Concepts and Skills **144**
Finding the New Amount **145**
Maintaining Concepts and Skills **146**
Calculating Sums and Differences **147**
Calculating Sums and Differences 148
Comparing Populations in Ohio **149**
Maintaining Concepts and Skills **150**
Using Data **151**
Maintaining Concepts and Skills **152**
Ordering 6-digit Numbers **153**
Ordering 6-digit Numbers 154
Working with One Million **155**
Adding On to Make One Million 156

Home Link page numbers are shown in gray.

Name Date

Maintaining Concepts and Skills

1. Ring the best unit of measure.

a. Height of a tree

inch foot mile

b. Width of a book

centimeter meter

2. Ted had 36 bottle caps. Dion gave him some more. Now he has 51 bottle caps. How many did Dion give him?

_______ bottle caps

3. Measure the lines.
Write the lengths.

This line is _______ inches long.

This line is _______ inches long.

4. What number is 100 more than 236?

____

What number is 20 more than 143?

____

5. Add in your head.

16 + 10 = _____ 18 + 18 = _____ 26 + 20 = _____

6. Subtract in your head.

14 – 5 = _____ 37 – 14 = _____ 47 – 13 = _____

Name Date

Counting On and Counting Back

Write what comes just **before** and just **after** each number.

_____ 70 _____	_____ 2,348 _____
_____ 89 _____	_____ 3,627 _____
_____ 110 _____	_____ 3,111 _____
_____ 129 _____	_____ 3,120 _____
_____ 349 _____	_____ 3,210 _____
_____ 500 _____	_____ 3,020 _____
_____ 560 _____	_____ 3,002 _____
_____ 699 _____	_____ 6,101 _____
_____ 704 _____	_____ 6,110 _____
_____ 710 _____	_____ 6,011 _____
_____ 990 _____	_____ 6,001 _____
_____ 1,234 _____	_____ 6,090 _____
_____ 1,480 _____	_____ 6,009 _____

Name Date

Maintaining Concepts and Skills

1. My giant step is 38 inches. How much is that in feet and inches?

_________ feet _________ inches

2. Measure the line. Write the length.

This line is _________ cm long.

3. Colin had 235 stickers. He gave 146 to Elsie. How many stickers did Colin have then?

_________ stickers

4. Gerard had 63 stickers. He gave some to Calvin. Then Gerard had 55 stickers. How many stickers did Gerard give to Calvin?

_________ stickers

5. a. Write the largest number you can using these digits:

2, 0, 7

____

b. Write the smallest number you can using these digits:

9, 3, 5

____

6. Write the number that has

3 tens, 4 ones

____

6 ones, 5 tens

____

Name Date

Counting On and Counting Back

Write what comes just **before** and just **after** each number.

_______ 120 _______	_______ 1,632 _______
_______ 131 _______	_______ 1,650 _______
_______ 199 _______	_______ 1,605 _______
_______ 201 _______	_______ 1,065 _______
_______ 370 _______	_______ 1,056 _______
_______ 399 _______	_______ 1,006 _______
_______ 610 _______	_______ 1,060 _______
_______ 649 _______	_______ 1,600 _______
_______ 711 _______	_______ 1,590 _______
_______ 799 _______	_______ 1,509 _______
_______ 890 _______	_______ 1,900 _______
_______ 989 _______	_______ 1,090 _______
_______ 999 _______	_______ 1,009 _______

Fourth-grade students should be confident about counting and sequencing numbers. On these pages, increasing and decreasing numbers by one encourages the students to think about place value and provides practice in counting on, counting back, and writing numbers.

Name Date

Maintaining Concepts and Skills

1. Answer these.

About how tall are you? ________

About how high is your classroom ceiling? ________

About how tall is a moving van? ________

2. Name something you would measure in

inches ________

feet ________

yards ________

3. Use these clues to write a number.

a. More tens than ones and fewer hundreds than ones.

____

b. 3 more tens than hundreds and 2 more ones than tens.

____

4. Write this number in words.

26

____

5. Add.

$$\begin{array}{r} 38 \\ +\ 4 \\ \hline \end{array} \qquad \begin{array}{r} 38 \\ +\ 14 \\ \hline \end{array} \qquad \begin{array}{r} 38 \\ +\ 24 \\ \hline \end{array}$$

6. Subtract.

$$\begin{array}{r} 72 \\ -\ 5 \\ \hline \end{array} \qquad \begin{array}{r} 72 \\ -\ 15 \\ \hline \end{array} \qquad \begin{array}{r} 72 \\ -\ 25 \\ \hline \end{array}$$

Name Date

Adding 1-digit and 2-digit Numbers

Add. Write the totals.

+	1	2	3	4
6	7			
16				
26			29	
7				
17				
27				

+	6	7	8	9
2				
12				
22				
5				
15				
25				

+	1	2	3	4
8				
18				
28				
9				
19				
29				

+	6	7	8	9
3				
13				
23				
4				
14				
24				

Name Date

Adding Amounts

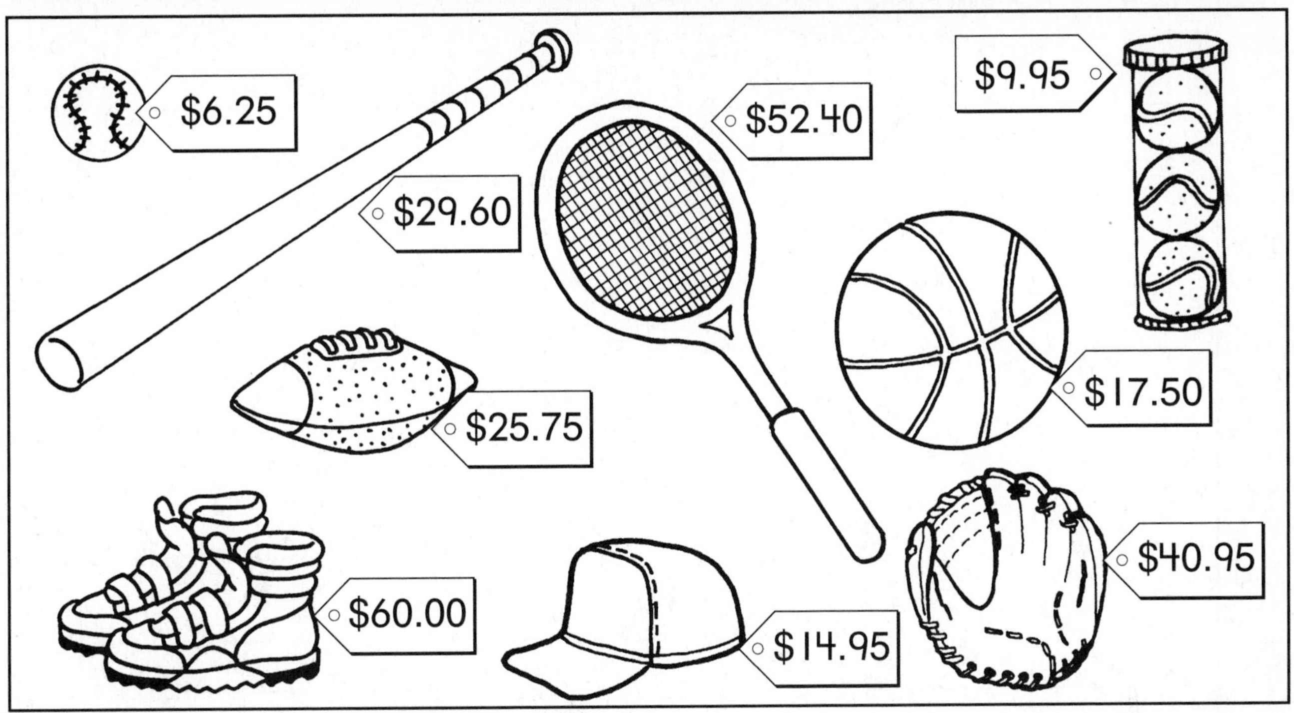

1. List 2 things you would buy with each gift certificate. No change is given. Choose things that give a total close to the amount on the certificate. (You can "buy" things more than once.)

2. Shade the bills you would use to pay for these.

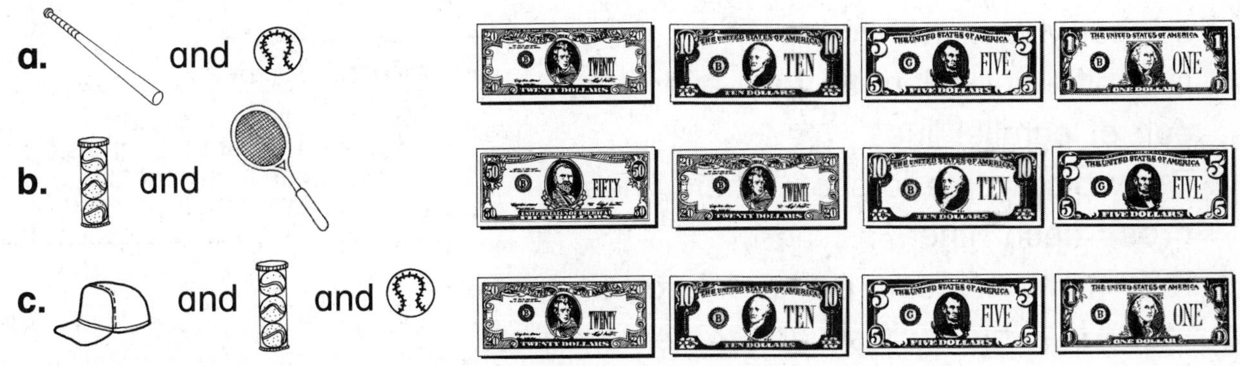

Name Date

Working with Parallel Lines

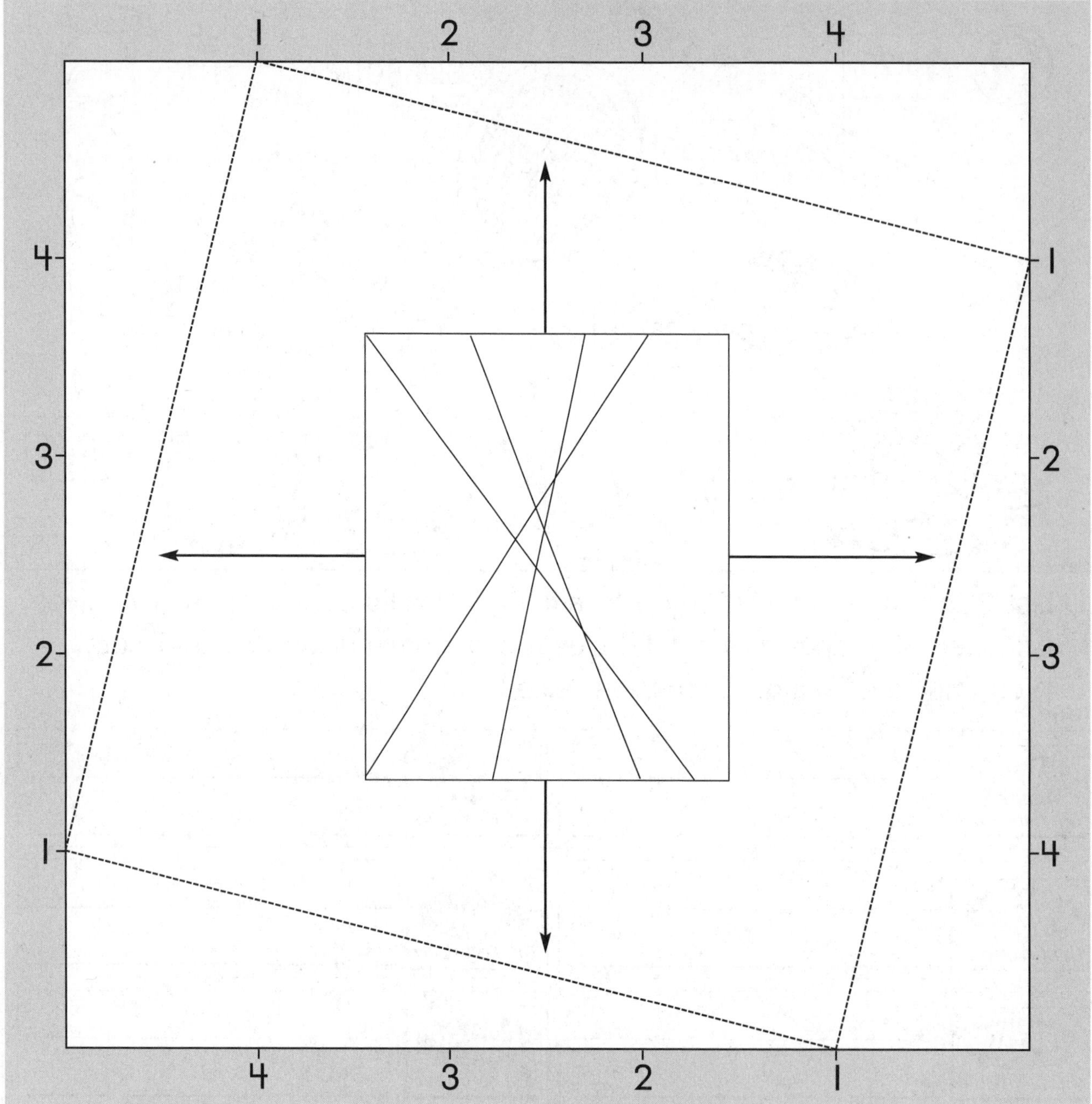

1. Draw lines to connect numbers that are the same, so that you make squares.

2. Use different colors to trace over each pair of parallel lines.

3. Draw black dots to show where lines **cross** each other.

4. How many crossing points did you find? ___________

5. Place a ✓ by each square corner you find.

Parallel Lines
Straight lines that are the same distance apart and never intersect are called parallel.

Name Date

Adding 1-digit and 2-digit Numbers

Add. Write the totals.

+	1	2	3	4
5				
15				
25				
35				
45				
55				

+	6	7	8	9
5				
15				
25				
35				
45				
55				

+	1	2	3	4
49				
59				
69				
79				
89				
99				

+	6	7	8	9
49				
59				
69				
79				
89				
99				

These charts review addition of 1-digit and 2-digit numbers. When your child has finished, ask him or her to describe the number patterns in the charts. Then ask your child to make up some other charts for further practice.

Name Date

Reading and Creating Bar Graphs

This graph shows the number of students at Riverview Elementary School.

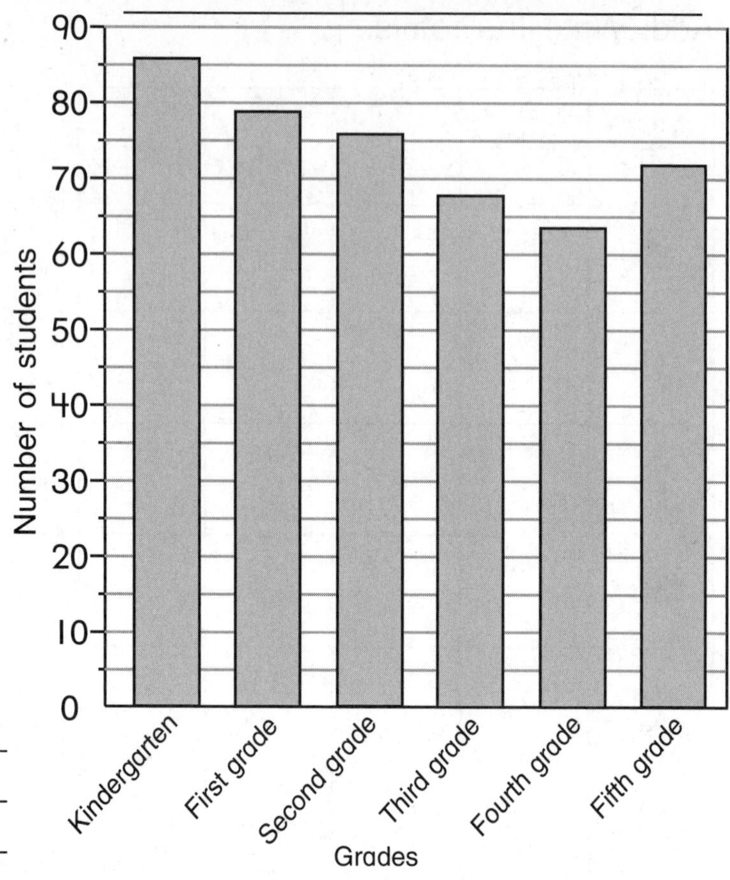

1. Write a title above the graph.
2. Which grade has the
   a. greatest number of students?

   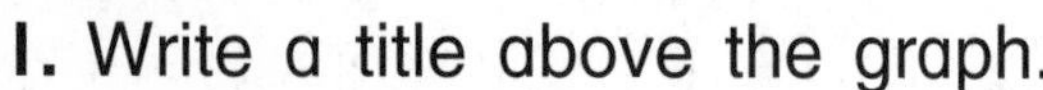

   b. least number of students?

   ____
3. Estimate how many **more** students are in kindergarten than
   a. First grade ______________
   b. Third grade ______________
   c. Fourth grade ______________
4. Which 2 grades have the **smallest** difference in numbers?

   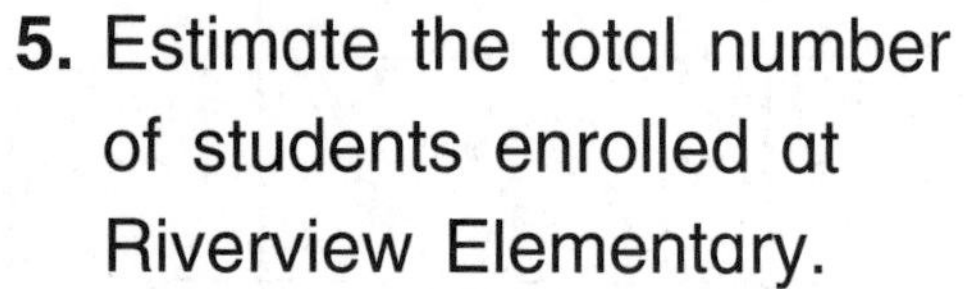

5. Estimate the total number of students enrolled at Riverview Elementary.

   ____
6. How many grades and students are at **your** school? Construct a graph to show the information.
7. Write a title for your graph.

Name Date

Maintaining Concepts and Skills

1. Name something you would measure in

   centimeters ________________

   meters ________________

2. Choose something bigger than a book and smaller than your classroom.

   What is it? ________________

   What is its perimeter? ____________

3. Ali had 83 fish. 47 were guppies and the rest were tetras. How many tetras did Ali have?

   ______________ tetras

4. Solve.
   Azura had 83 beads. Tanya had 17 more beads than Azura. How many beads did Tanya have?

   ________________ beads

5. Write the largest 3-digit number you can, using these clues.

   One digit is 5. Another digit is 3 less than the ones digit. The hundreds place is even.

   ____

   One digit is the sum of the other two. One digit is even.

   ____

6. Write a numeral to match each word.

   Seventy-six ____________

   One hundred thirteen ____________

   Two hundred six ____________

Name Date

Testing Triangles

1. Use 18 paper clips or toothpicks to construct all the different triangles you can make. Sketch a picture of each triangle you make in one of the boxes below. Label each side of every triangle.

a.

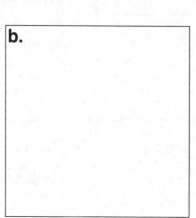

b.

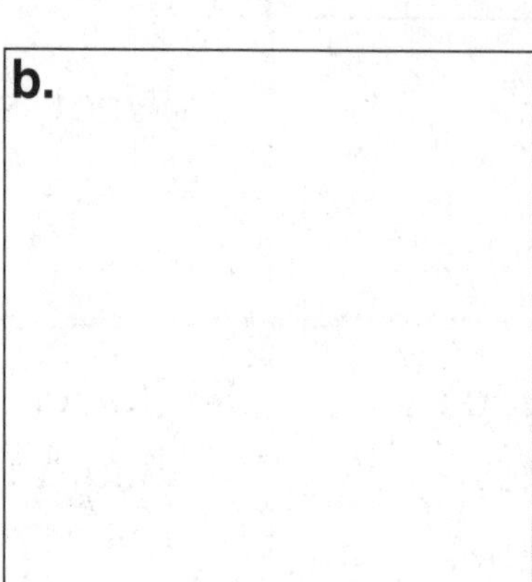

c.

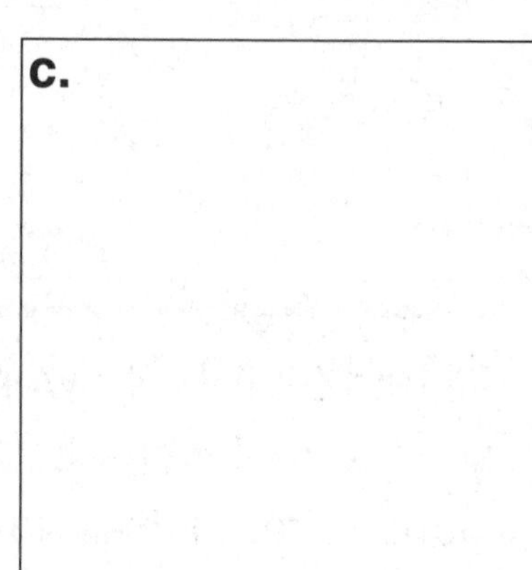

d.

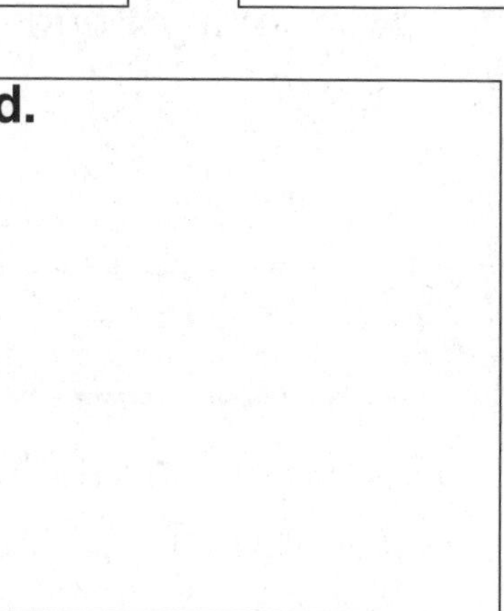

e.

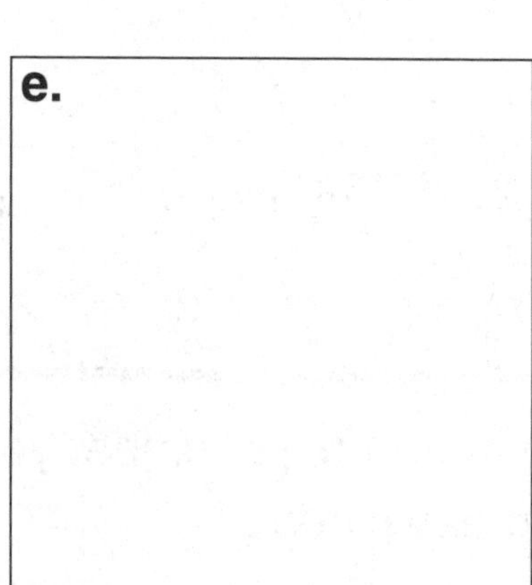

2. Now use the same 18 paper clips or toothpicks to try to create a triangle that has sides of 10, 5, and 3.

What did you discover? ______________________________

____

____

3. Write the length of the sides of 2 other triangles you CANNOT make with 18 paper clips or toothpicks.

____

____

Name Date

Maintaining Concepts and Skills

1. Luis had some pencils. His sister gave him 15 more. Now he has 41 pencils. How many pencils did Luis start with?

_________ pencils

2. Sofia had 35 coins in her collection. Alicia had 92 coins. How many more coins did Alicia have than Sofia?

___________ coins

3. Estimate, or use mental mathematics.

98 + 93 = ______

67 + ______ = 160

4. Write the answers.

26 + 38 = ______

17 + 9 − 8 + 14 = ______

5. Draw a triangle. Measure it with a centimeter ruler. Write its perimeter.

Perimeter _______________ cm

6. Draw a design that has 3 rectangles and 3 triangles.

Name Date

Relating Distances

Write the total distance. Then write whether the distance is **greater than**, **equal to**, or **less than** 1 kilometer.

600 m + 500 m = 1,100 m	greater than 1 km
750 m + 200 m = ________	________
450 m + 550 m = ________	________
475 m + 625 m = ________	________
3 × 400 m = ________	________
4 × 200 m = ________	________
300 m + 200 m + 200 m = ________	________
200 m + 400 m + 600 m = ________	________
250 m + 350 m + 450 m = ________	________
330 m + 340 m + 350 m = ________	________

Name Date

Maintaining Concepts and Skills

1. Use each digit once. 4, 3, 2, 1

Write the number that has

the greatest value __________

the least value __________

2. Use each digit once. 4, 5, 6, 7

Write a number

- less than 6,500 __________
- with 5 in the hundreds place __________

3. Use addition. Write a number sentence that has this sum.

____________________ = 415

4. Use subtraction. Write a number sentence that has this difference.

____________________ = 989

5. Solve.

Ron had $4.35. His uncle gave him $1.95. How much did Ron have then?

$__________

6. Solve.

411 + 135 − 390 − 119 + 13 − 50 = __________

Name Date

Reading and Creating Bar Graphs

Some Grade 4 students measured how far they could jump from a standing position. They constructed a graph to show their results.

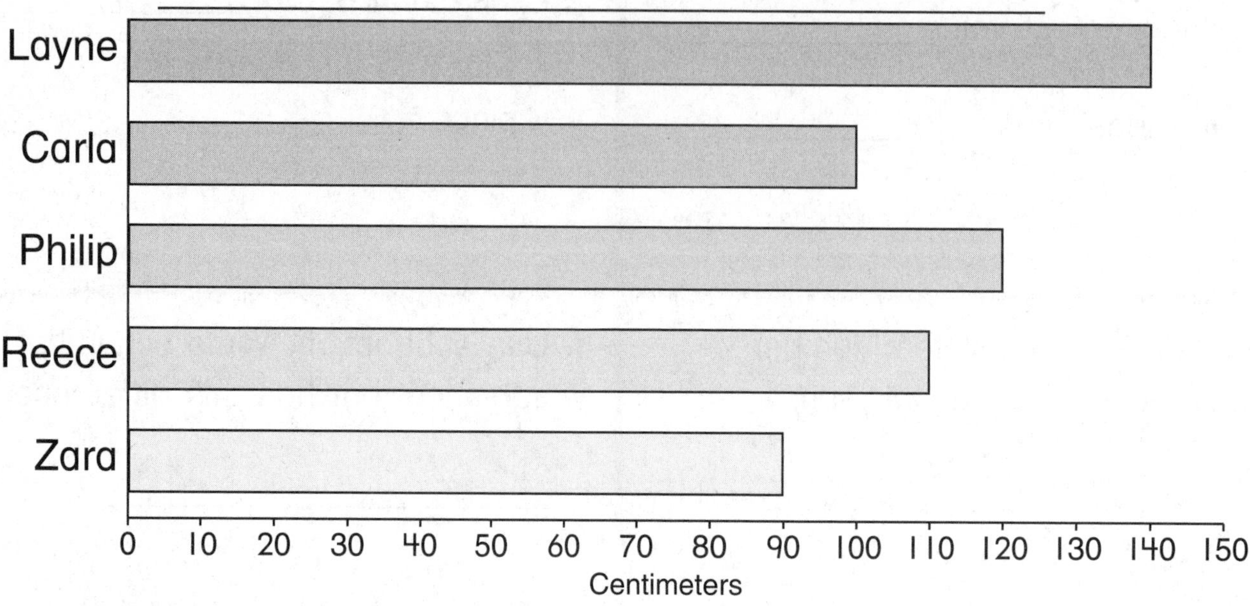

1. Write a title above the graph.
2. Who jumped the greatest distance? ____________
3. How much farther did Layne jump than Zara? ________ cm
4. Who jumped exactly 1 meter? ____________
5. Measure how far you can jump from a standing position. Then measure some friends' jumps. Construct a graph to show the results. Write a title for your graph.

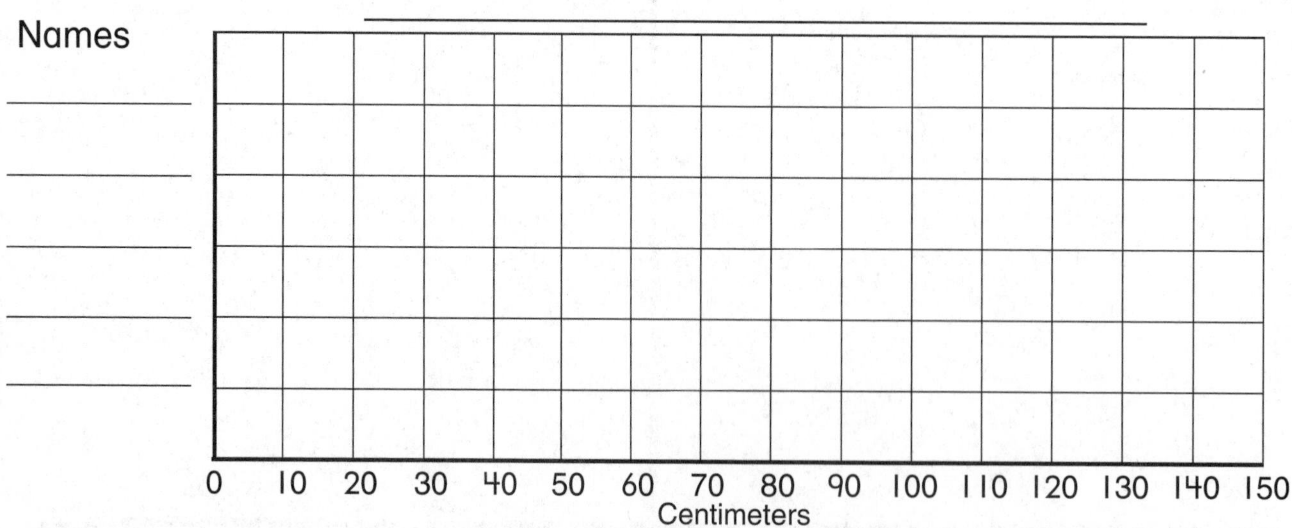

Name Date

Maintaining Concepts and Skills

1. Write 2 different **addition** problems that have a sum of 2,004.

_______________ = 2,004

_______________ = 2,004

2. Write 2 different **subtraction** problems that have a difference of 2,004.

_______________ = 2,004

_______________ = 2,004

3. Use | 3, 2, 9, 0 |

a. Make the largest number.

____

b. Make the smallest number.

____

4. Use | 5, 1, 9, 8 |

a. Make the largest number.

____

b. Make the smallest number.

____

c. What is the difference?

____

5. For each shape below, draw in all the lines of symmetry that you can find.

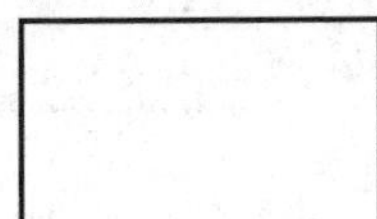

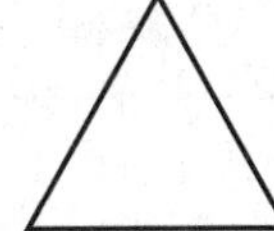

6. Draw a shape that has 8 sides.

What is it called? __________

Name Date

Using Millimeters

1. Write 2 things you know about millimeters.

____

____

2. Find things that are a **little less than** 1 millimeter thick and a **little more than** 1 millimeter thick.

A little less than 1 millimeter	A little more than 1 millimeter
______	______
______	______
______	______
______	______

3. Complete this chart.

______ millimeters is the same as 1 centimeter.

______ millimeters is the same as 10 centimeters.

______ millimeters is the same as 100 centimeters.

______ millimeters is the same as 1 meter.

4. What steps would you use to find the number of millimeters in 1 kilometer?

____

____

5. Choose a unit from the list to measure the lengths below.

millimeter centimeter meter kilometer

a. Your height ______

b. Length of a paper clip ______

c. Distance between your house and school ______

d. Distance around a tennis court ______

Name Date

Maintaining Concepts and Skills

1. Sal spent $43.85. When he got home, he had $12.95 left. How much did he start with?

$________

2. Carl sold José 137 minicars. Now José has 401 minicars. How many minicars did José start with?

________ minicars

3. Add.

58 + 6 = ______

58 + 16 = ______

4. Subtract.

33 − 15 = ______

43 − 15 = ______

5. Using each of the shapes in the box, draw a shape that is worth $6.95.

triangle: 30¢
square: $1.25
rectangle: $1.00

Label the value of each shape.

6. Solve.

I have no digits the same.
My thousands digit is less than my tens digit.
My hundreds digit is twice my ones digit.
If you add my digits, the sum is 25.
Who could I be? ______________
Could I be 2 numbers? ______________

Name Date

Relating Distances

Write the total distance. Then write whether the distance is **greater than**, **equal to**, or **less than** 1 kilometer.

550 m + 400 m = ___________	___________
400 m + 700 m = ___________	___________
300 m + 300 m + 300 m = ___________	___________
200 m + 300 m + 500 m = ___________	___________
2 × 550 m = ___________	___________
4 × 220 m = ___________	___________
700 m + 100 m + 300 m = ___________	___________
150 m + 350 m + 550 m = ___________	___________
3 × 330 m = ___________	___________
4 × 250 m = ___________	___________

It is essential that students are familiar with metric units, which are used in science, higher mathematics, and medicine, including nursing. Your child may work mentally or use paper and pencil to figure out the total distances. He or she should know that there are 1,000 meters in a kilometer.

Name Date

Reading Numbers on a Number Line

Write the numbers the arrows are pointing to.

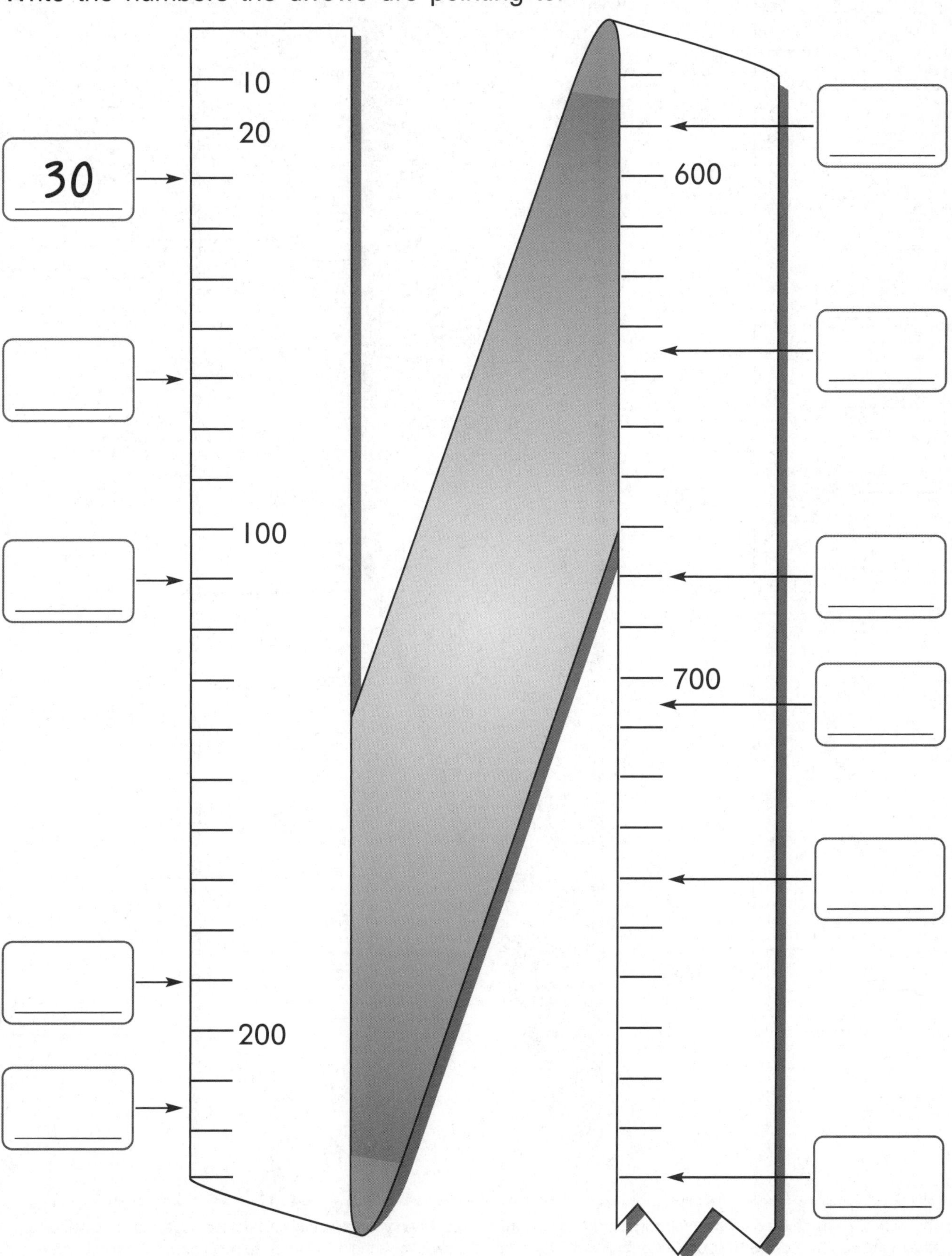

Name Date

Reading Numbers on a Number Line

Write the numbers the arrows are pointing to.

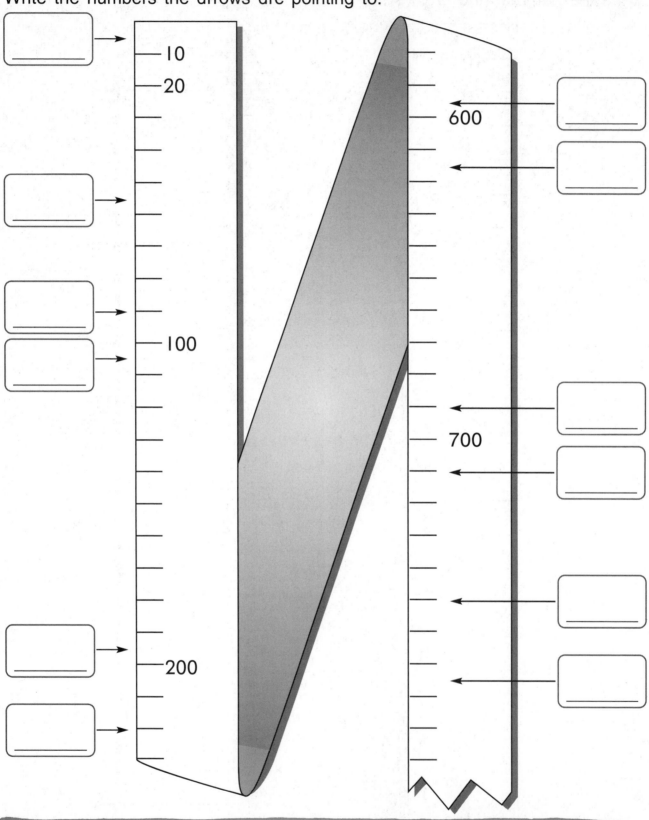

This activity helps develop skills for reading rulers and tape measures. When your child has finished the page, say or write some other numbers, such as 45, 150, and 675. Challenge him or her to draw arrows that point to those numbers on the number line.

Name Date

Maintaining Concepts and Skills

1. Cindy bought 7 gifts. Each gift cost the same amount. She spent $42.00. How much did each gift cost?

   $__________

2. Fred Frog can jump 6 times his length. Fred is 9 cm long. How long is Fred's jump?

   __________ cm

3. Two children shared 5 cookies. How many cookies did each child get?

   ______ cookies

4. Which is more, $\frac{1}{2}$ or $\frac{1}{3}$ of the same thing? ______________

   How do you know?

   ____

   ____

   ____

5. Continue these patterns.

   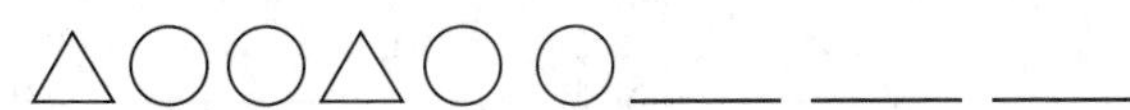 ___ ___ ___

   a, c, e, ___, ___, ___

   12, 17, 22, 27, ____, ____, ____

6. Mark this shape to show thirds.

Name Date

Using Addition, Subtraction, Multiplication, and Division

1. Draw lines to match each problem to the correct story. Write the answer.

18×6	Ed has saved $6. He wants to buy a CD that costs $18. How much more money does he need to save?
$18 + 6$	Each car on a train is 18 yards long. If a train has 6 cars, how long is the train?
$18 - 6$	We have 18 tennis balls and 6 containers. How many tennis balls can go in each container so they each hold the same number?
$6\overline{)18}$	18 children were on the school bus. 6 more children got on. How many children are on the bus now?

2. Write an addition and a subtraction story for these two price tags.

$17

$9

____

____

____

____

3. Write a multiplication and a division story for this picture.

$15

____

____

____

____

Name Date

Investigating Distances

Plains County Road Distances

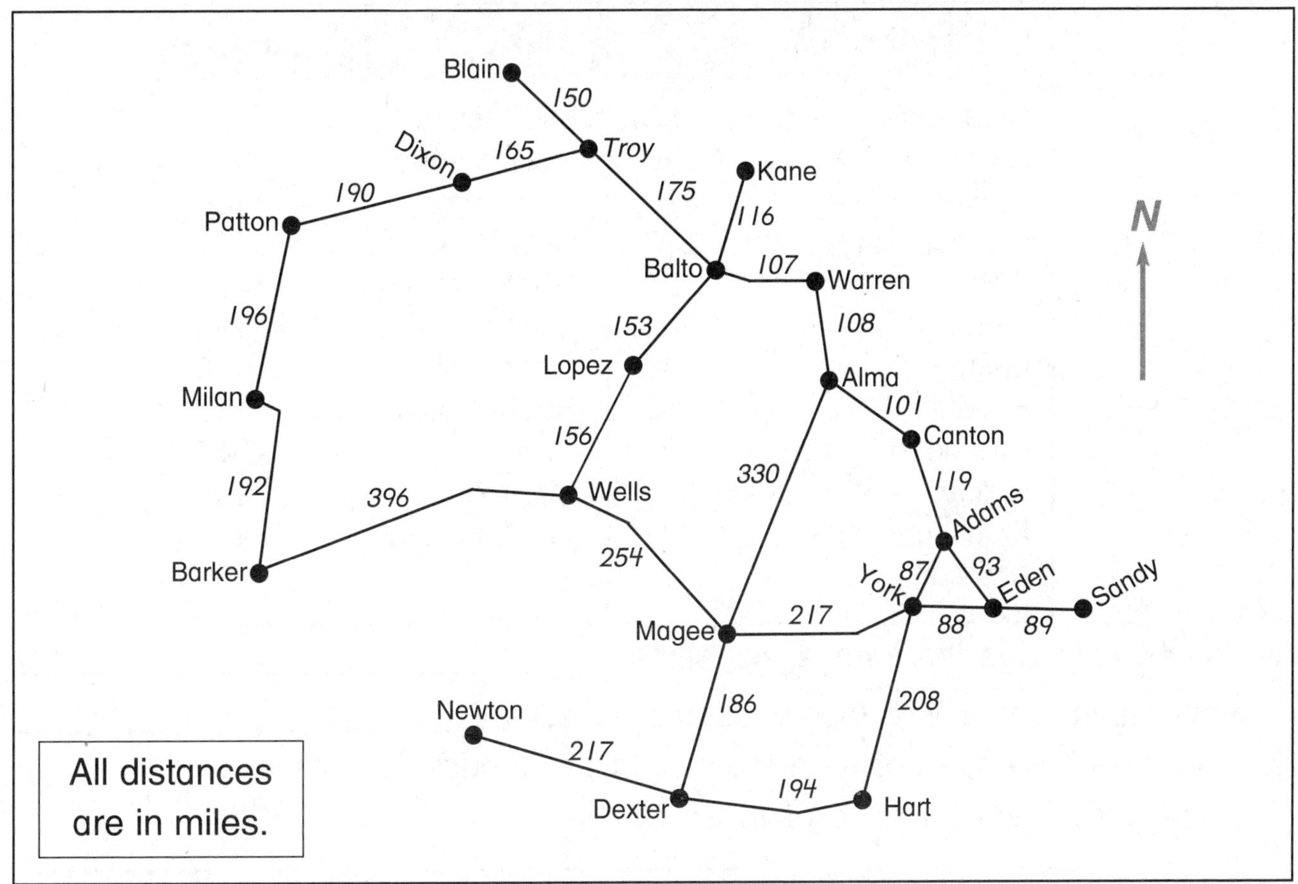

1. How much farther is Wells to Barker than Wells to Balto? ________ miles

2. How far would you travel if you drove the shortest route from
   a. Troy to Patton? ________________ miles
   b. Alma to Dexter? ________________ miles

3. Calculate the **shortest** road distances between these towns.
   a. York and Dexter ________________ miles
   b. Balto and Magee ________________ miles

4. A bus driver wants to drive from Sandy to Troy. Calculate the length of the shortest trip. ________________ miles
   Name the towns that the bus will drive through.
   Sandy ➡ ____

   ____

Name Date

Comparing Coastlines

This chart gives the coastline length for every state that has a coastline.

State coastline (miles)		State coastline (miles)	
Alabama	53	Mississippi	44
Alaska	6,640	New Hampshire	13
California	830	New Jersey	130
Connecticut	253	New York	127
Delaware	28	North Carolina	301
Florida	1,350	Oregon	296
Georgia	100	Rhode Island	40
Hawaii	750	South Carolina	187
Louisiana	397	Texas	367
Maine	228	Virginia	112
Massachusetts	192	Washington	157
Maryland	31		

1. Which state has the longest coastline? ____________________

   How much longer is it than Florida's coastline? ____________________

2. Find the difference between the length of Florida's coastline and the coastline of each of these states.

	Delaware	Hawaii	Maine	New York	Texas
Difference	______	______	______	______	______

3. Calculate the total length of these coastlines.

   a. Washington, Oregon, and California ______________

   b. Texas, Louisiana, Mississippi, and Alabama ______________

   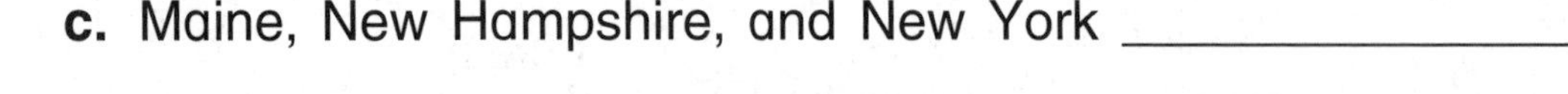

   c. Maine, New Hampshire, and New York ______________

4. Calculate the combined length of these coastlines.

   a. Alaska and Hawaii ______________

   b. All the other states together ______________

   c. Which combined coastline is longer? ____________________

5. Calculate the total length of the United States coastline. ______________

Name Date

Adding and Subtracting 3-digit Numbers

Add or subtract to find the missing number.
Write the answer.

You might need to use the calculating space below.

360 + 250 = ___________

480 − 390 = ___________

250 + ___________ = 470

___________ + 375 = 465

___________ = 275 + 150

___________ = 865 − 220

680 − ___________ = 360

___________ − 245 = 350

Calculations:

Name Date

Addition and Subtraction Problems

This chart shows how much different classes raised by selling magazines.

Grade 3		Grade 4		Grade 5		Grade 6	
Room 5	Room 6	Room 7	Room 8	Room 9	Room 10	Room 11	Room 12
$1,975	$1,256	$3,064	$2,119	$2,800	$2,104	$3,085	$1,046

1. Calculate how much money each grade raised.

Grade 3	Grade 4	Grade 5	Grade 6
________	________	________	________

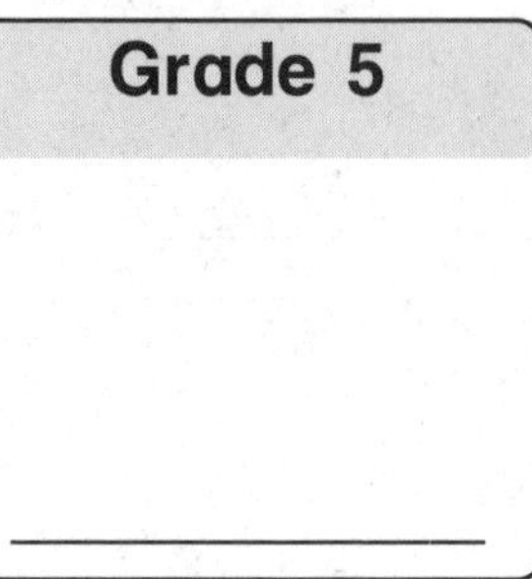

2. Which **grade** raised the most? __________ The least? __________

   Calculate the difference. ______________

3. How much more money did Grade 5 raise than Grade 6? __________
   Show how you figured it out.

4. Calculate the differences.

Rm. 5 and Rm. 6	Rm. 7 and Rm. 8	Rm. 9 and Rm. 10	Rm. 11 and Rm. 12
________	________	________	________

5. Each room's students wanted to raise $2,500. List the rooms where they reached that goal. Then calculate how much more than $2,500 those students raised.

________________ ________________ ________________

________________ ________________ ________________

Name Date

Maintaining Concepts and Skills

1. Ring the 2 numbers that will give a product closest to the target number.

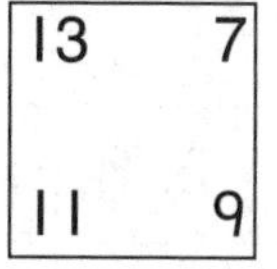

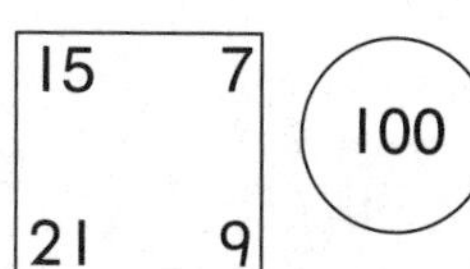

2. Use mental math to find the answers.

$3 \times 31 =$ ______

$5 \times 21 =$ ______

$9 \times 13 =$ ______

3. In each box, ring the fraction that is more.

$\frac{1}{2}$ $\frac{1}{4}$

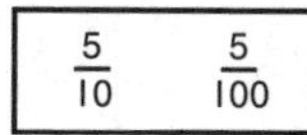

How did you decide?

____

____

____

4. Ed ate $2\frac{1}{2}$ muffins. Hollie ate $1\frac{1}{2}$ muffins. How many muffins did they eat in all?

________ muffins

5. Start at 0. Count by halves until you get to 5. How many halves did you count?

____

6. Write a story problem that uses multiplication.

____

____

____

____

____

Name Date

Subtracting 3-digit Numbers

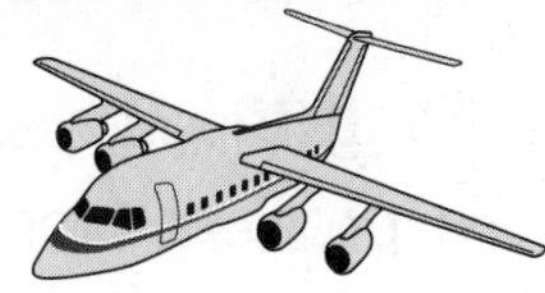

Subtract to find the number of empty seats on each flight.

	Flight 1	Flight 2	Flight 3	Flight 4
Seats available	285	264	328	265
Passengers	139	145	162	194
Empty seats				

	Flight 5	Flight 6	Flight 7	Flight 8
Seats available	134	232	416	325
Passengers	86	165	288	169
Empty seats				

	Flight 9	Flight 10	Flight 11	Flight 12
Seats available	405	330	260	208
Passengers	128	168	185	179
Empty seats				

Use with Investigation 3.3

Name Date

Maintaining Concepts and Skills

1. Write 3 different multiplication problems that have a product of 75.

______

______

______

2. Write a story problem for one of your multiplication problems.

______

______

______

______

______

3. Darrin bought 3 shirts and 2 pairs of shorts. How much did he spend?

$______

4. Continue these patterns.

1, 5, 9, 13, 17, ____, ____, ____

100, 94, 88, 82, ____, ____, ____

2, 1, 4, 3, 6, 5, ____, ____, ____

5. Mark and shade $\frac{3}{5}$ of the rectangle.

6. Mark where you think $\frac{2}{3}$ and $\frac{1}{10}$ would be on the number line.

0 1

Name Date

Subtracting 3-digit Numbers

Subtract to find the number of empty seats on each flight.

	Flight 1	Flight 2	Flight 3	Flight 4
Seats available	362	48	327	415
Passengers	225	29	144	263
Empty seats				

	Flight 5	Flight 6	Flight 7	Flight 8
Seats available	215	346	422	322
Passengers	166	189	287	266
Empty seats				

	Flight 9	Flight 10	Flight 11	Flight 12
Seats available	405	380	305	280
Passengers	256	195	288	176
Empty seats				

Your child may use any method he or she wishes to solve these problems. The examples have been carefully sequenced so that children who use a subtraction algorithm practice "one renaming" before "two renamings" and "renaming a zero."

Name Date

Maintaining Concepts and Skills

1. Al had 3 times as many coins as Dave. Dave had 35 coins. How many coins did Al have?

__________ coins

2. Each box holds 12 pencils. How many boxes are needed to hold 60 pencils? __________

How did you decide?

____

____

3. Write multiplication problems using these numbers. Use all the numbers.

31	93	6	120	3	20

____

____

4. Mark and shade the squares to show that $\frac{1}{2}$ is more than $\frac{1}{3}$.

5. Write a fraction that is close to one whole.

____

How do you know?

____

____

____

6. Write a fraction that is less than $\frac{1}{2}$.

____

How do you know?

____

____

____

Name Date

Addition and Subtraction Problems

These pictures show types of planes and the number of passengers they carry.

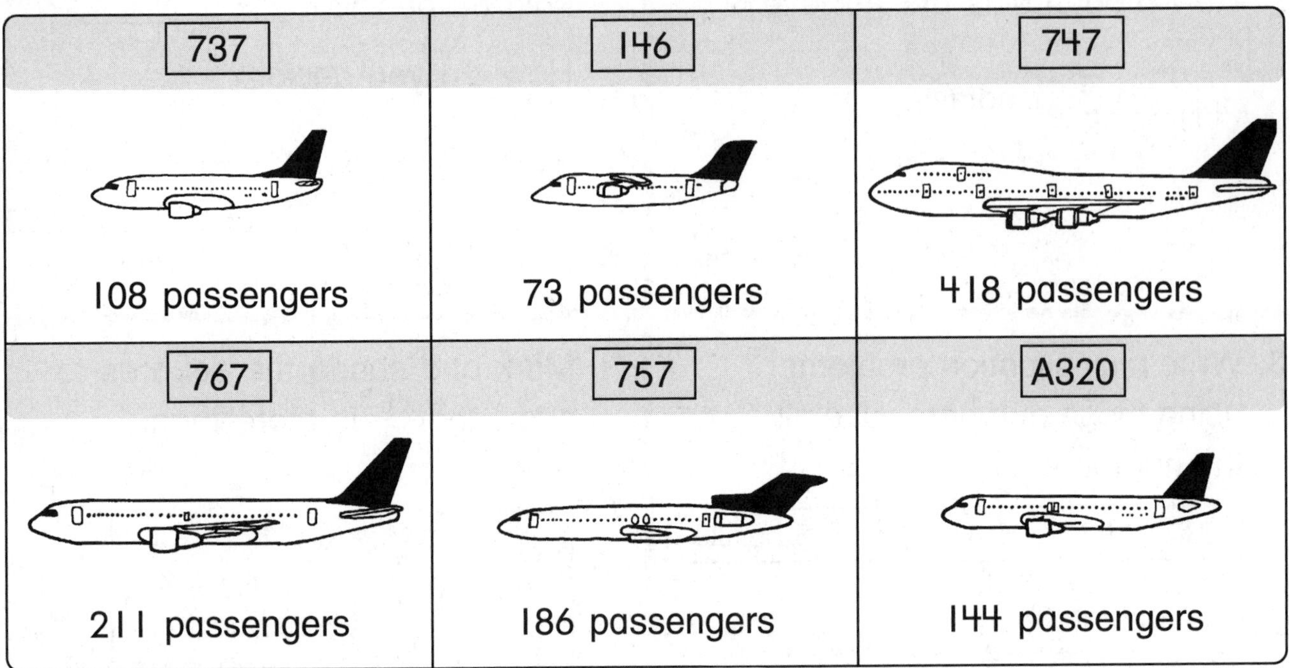

737	146	747
108 passengers	73 passengers	418 passengers
767	**757**	**A320**
211 passengers	186 passengers	144 passengers

1. How many more passengers does the 757 carry than the A320? ________ passengers

   a. Which plane holds the most passengers? ________________

   b. Which plane holds the fewest passengers? ________________

   c. What is the **difference** between the number of passengers these 2 planes carry? ________ passengers

2. How many more passengers would a 757 need to carry to hold the same number of passengers as the 747 ?

   ________ passengers

3. Write the number of students in your school. ________

   Write the number of passengers a 747 can carry. ________

   Calculate the difference. ________

4. Which plane would be needed to carry every student in your school? (Leave as few empty seats as possible.) ________

Name Date

Using a Number Expander

1. Read the number and **write** it on the number expander.

- six thousand, four hundred eighty-one

	thousands		hundreds		tens		ones

- four thousand, nine hundred fifty-three

	thousands		hundreds		tens		ones

- one thousand, seven hundred seventy-two

	thousands		hundreds		tens		ones

2. Read each number expander. **Write** the number in words.

5	thousands	6	hundreds	9	tens	2	ones

____

7	thousands	5	hundreds	7	tens	4	ones

____

3	thousands	8	hundreds	6	tens	1	ones

____

3. Draw lines to link each number with the matching number in words.

7,935	three thousand, nine hundred seventy-five
9,753	seven thousand, nine hundred thirty-five
3,975	nine thousand, seven hundred fifty-three
9,357	nine thousand, three hundred fifty-seven

4. Read the number in words. **Write** the number.

five thousand, eight hundred eighty-nine ______________________

two thousand, one hundred forty-six ______________________

seven thousand, two hundred thirty-one ______________________

Name Date

Creating 4-digit Numbers

These digit cards were drawn from a deck.

9	3	1	8

1. Use each digit **once** to make
   - the largest number possible ________________
   - the smallest number possible ________________
   - the largest **even** number possible ________________
   - the smallest **odd** number possible ________________
   - the numbers that are between 8,000 and 8,300

   ____

These are the cards from a second draw.

5	4	0	9

2. Use each digit once to make
   - the largest number possible ________________
   - the smallest number possible ________________
   - the largest **odd** number possible ________________
   - the smallest **even** number possible ________________
   - a number that is close to **5,000** ________________
   - a number that is close to **9,000** ________________
   - the **3** smallest numbers possible

   ____
   - the **3** largest numbers possible

   ____

Name Date

Maintaining Concepts and Skills

1. Draw hands on this clock to show 12:15.

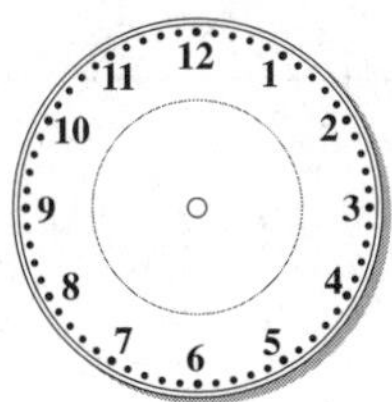

2. If the sun rose at 6:30 a.m. and set at 5:30 p.m., how many hours was the sun up?

____

How did you get your answer?

____

____

____

3. Four children shared 10 cookies. How many cookies did each child get?

_______ cookies

4. Frieda Frog can jump 6 times her length. Frieda is 12 cm long. How long is Frieda's jump?

__________ cm

5. Draw 2 shapes that are symmetrical. Mark the lines of symmetry.

6. Draw a rectangle that is 2 cm high and 3 cm long. What is its perimeter?

Perimeter __________ cm

Name Date

Reading and Writing 4-digit Numbers

Write the number.

one thousand, six hundred forty ______

four hundred sixteen ______

four thousand sixty ______

six thousand fourteen ______

one thousand, four hundred six ______

four thousand, one hundred sixteen ______

six thousand, one hundred four ______

one thousand sixty-four ______

four thousand six ______

six thousand four ______

Write the number in words.

3,012 ______

2,003 ______

3,020 ______

1,320 ______

3,102 ______

1,003 ______

1,203 ______

2,013 ______

2,310 ______

3,002 ______

Name Date

Writing About Thousands

1. Write 2 things you know about ten thousand.

____

____

2. Write where you might see ten thousand things in your

home _______________

city park _______________

town _______________

3. How far would you walk if you

walked 10,000 meters? __________ kilometers

walked 10,000 centimeters? __________ meters

4. Estimate the height of a stack of 10,000 quarters. __________

How could you check? _______________

____

5. How many of these would you need to make ten thousand?

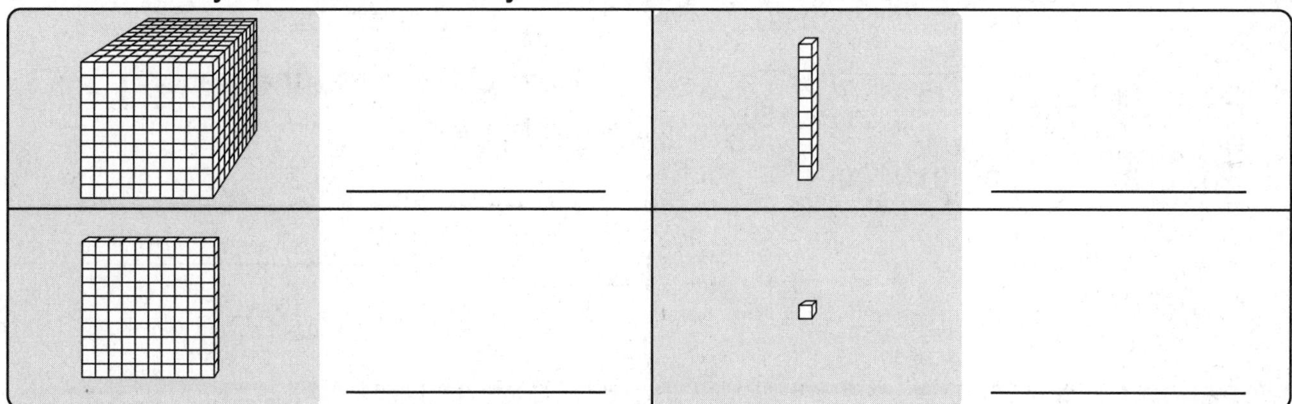

6. Show 5 different ways of writing ten thousand.

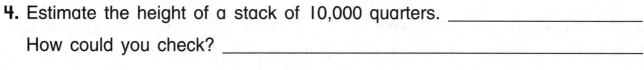

____ ten thousand	____ thousands	____ hundreds
____ tens	____ ones	

Name Date

Maintaining Concepts and Skills

1. Darcy got up at 7:15 a.m. She left for school at 8:38 a.m. Show the time that she got up and the time that she left.

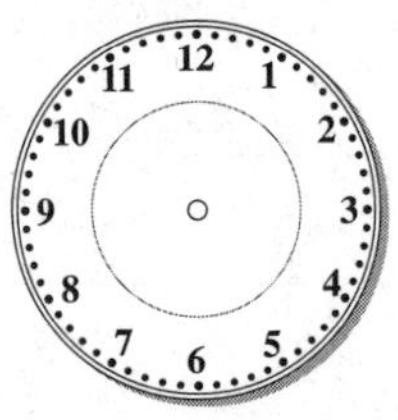
Got up

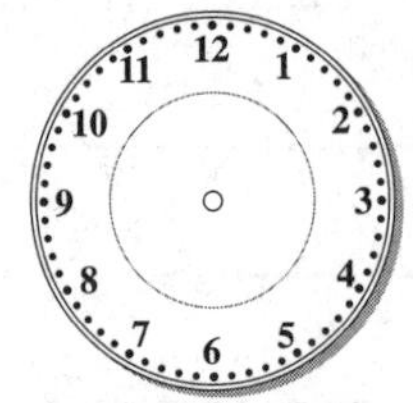
Left for school

2. Tom left for school at 8:00 a.m. It took him 10 minutes to walk to the bus stop. He rode the bus for 10 minutes. He talked to his friends outside school for 10 minutes. What time did he go into school?

____

Write a number sentence to show how you solved the problem.

____

3. In each box, ring 2 numbers that give a product closest to the target number.

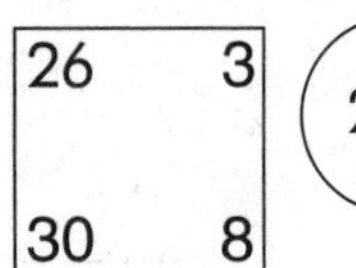

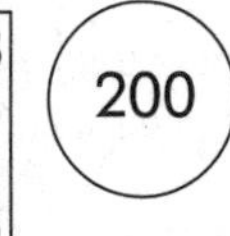

15	7
21	9

150

4. Use mental math.

$4 \times 26 =$ _____ $3 \times 29 =$ _____

$9 \times 15 =$ _____

How did you do the second problem?

____

____

5. Write a story for this number sentence.

$28 \div 4 =$ _____

____

____

____

6. Solve mentally.

$3 \times 39 =$ _____

How did you do it?

____

____

____

Name Date

Using a Number Expander

1. Read each number name. Write each number on the expander below.

twenty-six thousand, nine hundred thirty-one

eighty-two thousand, six hundred forty-five

seventy-three thousand, four hundred ninety-nine

2. Read each number name. Write the number beside it.

forty-two thousand, five hundred eighty-three __________

thirty-two thousand, six hundred seventy-five __________

twenty-seven thousand, one hundred fifty-nine __________

sixty-nine thousand, two hundred forty-one __________

3. Read each number expander. Write the number in words below it.

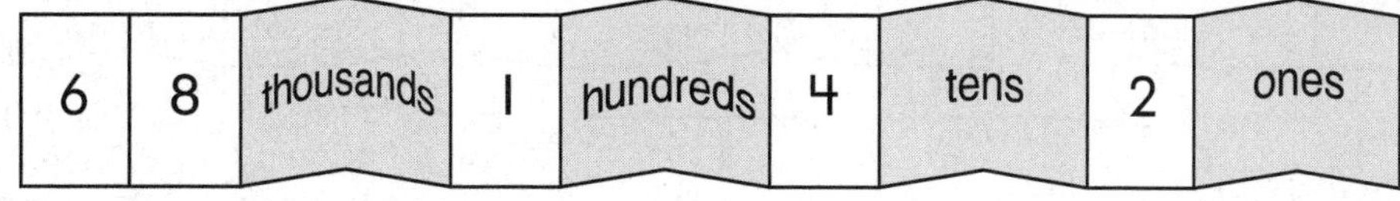

____

____

4. Write each number in words.

51627 ______________________________

39184 ______________________________

Name Date

Completing a Number Expander

Fill in the missing parts to finish each mix-and-match puzzle.

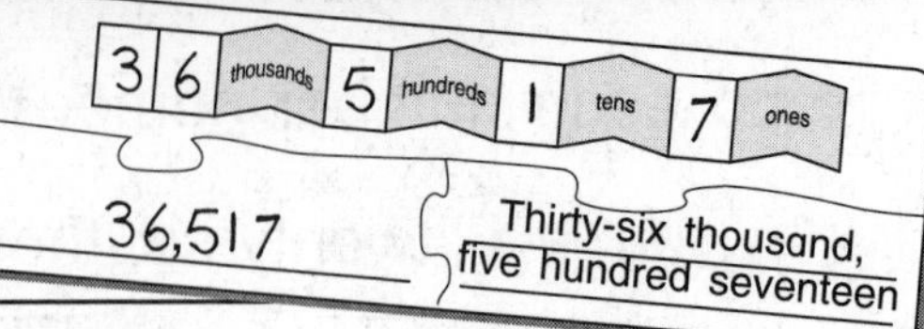

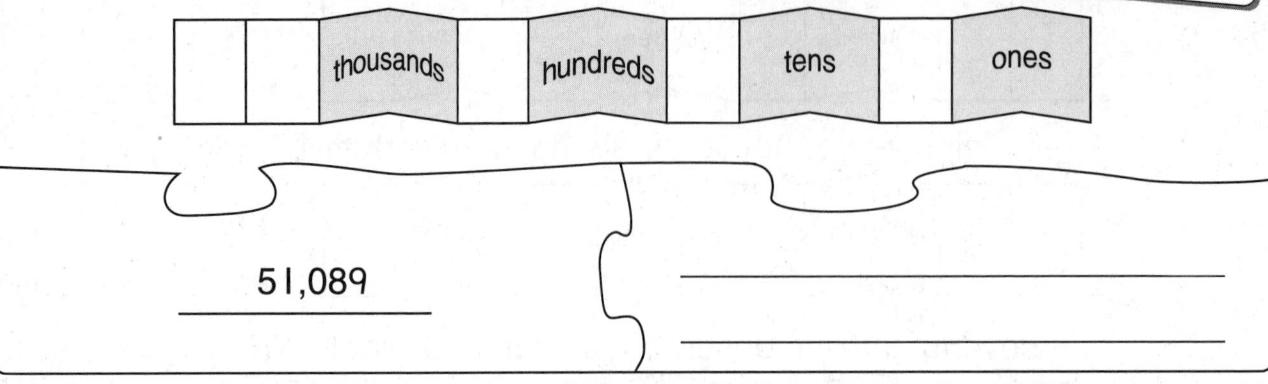

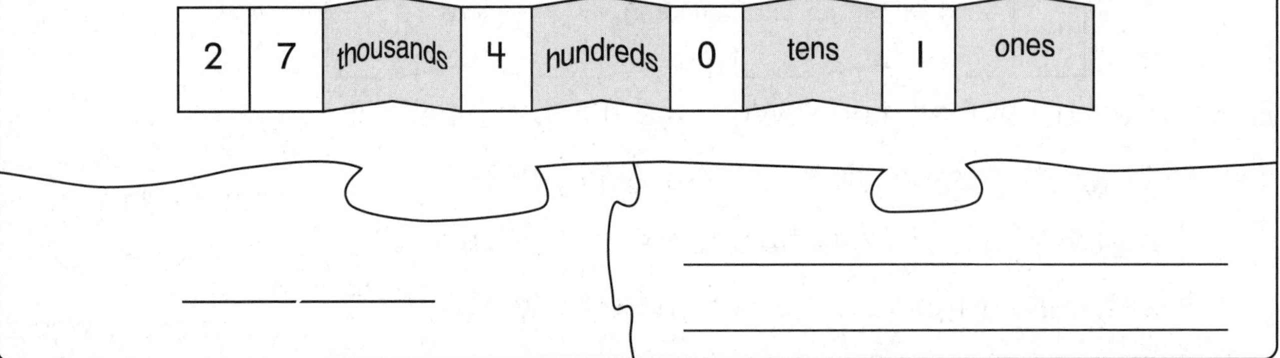

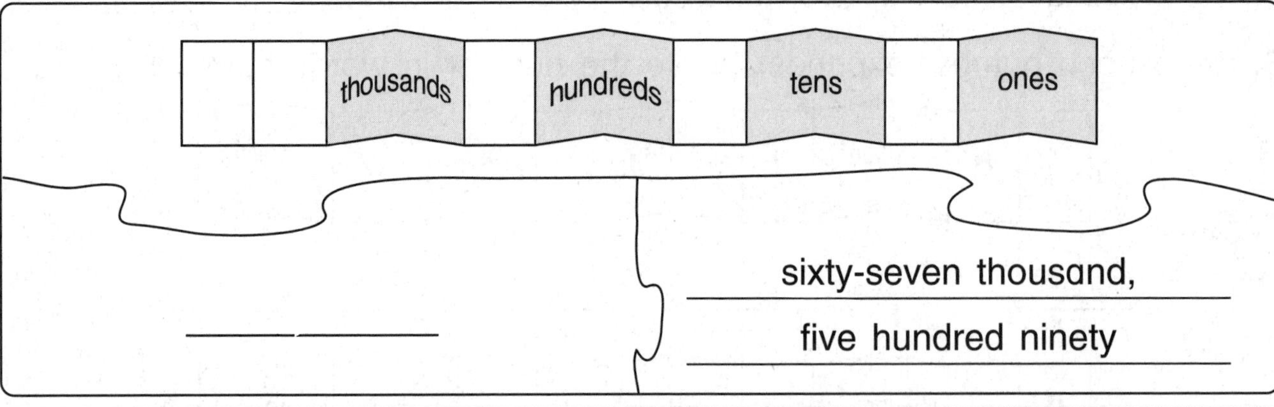

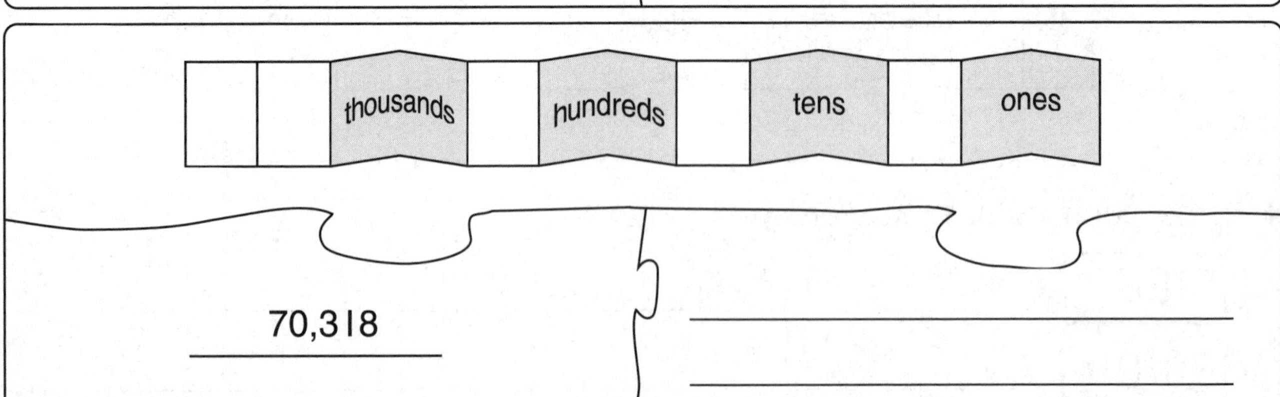

Name Date

Writing 5-digit Numbers

Each number below was made from these 5 digits.

6	3	1	8	2

Write each number in words.

31,826 ______________________________

____

82,613 ______________________________

____

26,138 ______________________________

____

16,823 ______________________________

____

Use the same 5 digits to write 4 different numbers.
Then write each number in words.

________ ______________________________

____

________ ______________________________

____

________ ______________________________

____

________ ______________________________

____

Name Date

Comparing 5-digit Numbers

Write **is less than** or **is greater than** to make each sentence true.

36,428 ___is less than___ 37,650.

87,009 ______________ 59,006.

21,613 ______________ 20,613.

42,631 ______________ 42,361.

12,401 ______________ 21,401.

31,760 ______________ 30,991.

< means
"is less than."
> means
"is greater than."

Write the symbol < or > to make each number sentence true.

51,261 __<__ 51,621	16,251 ______ 16,521
41,256 ______ 41,246	30,006 ______ 30,060
17,707 ______ 17,720	17,211 ______ 17,212
22,210 ______ 22,120	42,710 ______ 42,701
61,010 ______ 61,100	30,012 ______ 30,021

Name Date

HOME LINK

Comparing 5-digit Numbers

Write **is less than** or **is greater than** to make each sentence true.

17,612 ___is greater than___ 17,261.

59,071 ______________________ 59,901.

52,990 ______________________ 52,909.

37,619 ______________________ 37,616.

44,215 ______________________ 45,412.

19,019 ______________________ 19,109.

Write the symbol < or > to make each number sentence true.

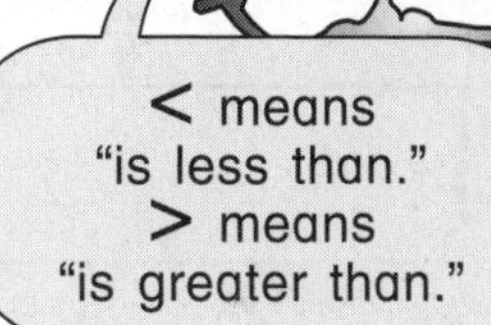
< means
"is less than."
> means
"is greater than."

71,717 __>__ 70,707	36,019 _____ 36,910
19,099 _____ 19,900	41,114 _____ 41,104
28,799 _____ 28,979	11,101 _____ 11,011
31,909 _____ 31,990	29,999 _____ 30,001
27,564 _____ 27,560	42,700 _____ 42,698

Sequencing numbers is an important skill. We often need to compare numbers and decide which is larger. These pages provide practice in comparing pairs of 5-digit numbers. A good way for your child to remember which symbol is which is to think of them as arrows that always point to the smaller number.

Name Date

Maintaining Concepts and Skills

1. Show the times on the clocks.

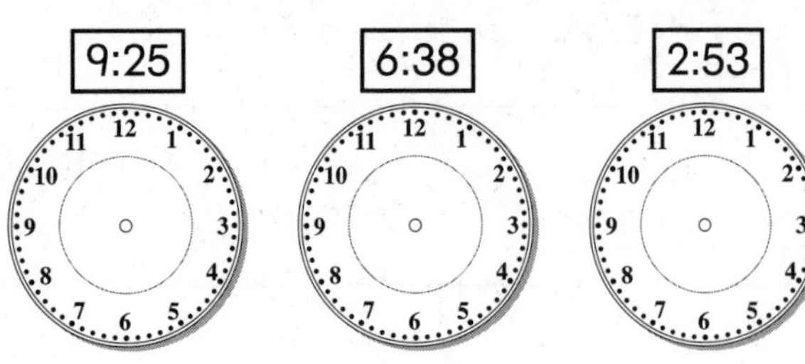

2. The plane departed at 7:35 a.m. The flight took 3 hours and 10 minutes. What time did the plane arrive? ______________

Show the times that the plane departed and arrived.

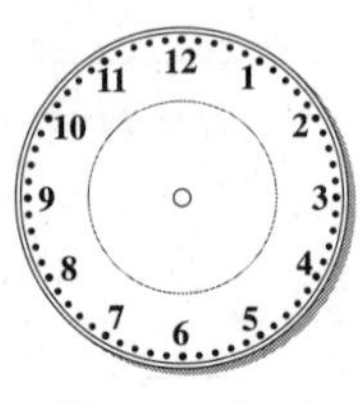
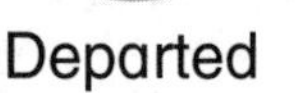
Departed

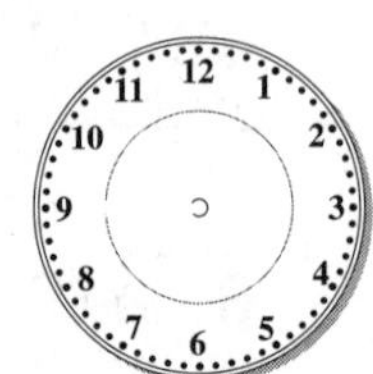
Arrived

3. Mark and shade $\frac{3}{8}$ of the rectangle.

4. Mark where you think $\frac{7}{8}$ and $\frac{3}{10}$ would be on the number line.

0 1

5. The class collected paper for recycling. They averaged 7 pounds per student. There are 23 class members. How many pounds of paper did they collect?

______ pounds

6. Write the answers.

$11 \times 14 =$ ________

$8 \times 48 =$ ________

$6 \times 62 =$ ________

Name Date

Increasing and Decreasing Numbers

Write the missing numbers.

	ten thousand more	one thousand more	one hundred more	ten more	one more
Starting Number	49,919	49,919	49,919	49,919	49,919
	ten thousand less	one thousand less	one hundred less	ten less	one less

	ten thousand more	one thousand more	one hundred more	ten more	one more
Starting Number	59,990	59,990	59,990	59,990	59,990
	ten thousand less	one thousand less	one hundred less	ten less	one less

	ten thousand more	one thousand more	one hundred more	ten more	one more
Starting Number	19,009	19,009	19,009	19,009	19,009
	ten thousand less	one thousand less	one hundred less	ten less	one less

This activity focuses on place value and sequencing skills. Before your child writes each answer, ask him or her to say which digit or digits in the starting number will be changed.

Name Date

Increasing and Decreasing Numbers

1. The people who had these tickets won the 4 major prizes at a football game.

BIG CITY STADIUM Ticket Number 97,629	BIG CITY STADIUM Ticket Number 68,001	BIG CITY STADIUM Ticket Number 85,999	BIG CITY STADIUM Ticket Number 60,019

The people who had a number that was **one more** than or **one less** than each winning ticket number were given a free ticket.
What were the numbers on their tickets?

	Winning ticket 97,629	**Winning ticket** 68,001	**Winning ticket** 85,999	**Winning ticket** 60,019
One more				
One less				

2. Read the headings. Fill in the missing numbers.

$1,000 less	$100 less	$10 less	Starting price	$10 more	$100 more	$1,000 more
			$36,250			
			$47,935			
			$79,641			
			$80,327			
			$42,059			
			$57,096			
			$85,110			
			$69,403			
			$51,076			
			$98,320			

Name Date

Maintaining Concepts and Skills

1. Jon had to take medicine every $1\frac{1}{2}$ hours. He started taking it at 8:00 a.m. At what times would he take it up until his bedtime at 10:00 p.m?

____

____

____

2. It takes 1 hour and 20 minutes for Don to travel to and from work each day. How much time does he spend traveling each week?

_______ hours _______ minutes

3. Write a fraction that is less than $\frac{1}{4}$.

____

How do you know it is less?

____

____

4. In each box, ring the fraction that is larger.

$\frac{5}{2}$	$\frac{5}{4}$

$\frac{2}{5}$	$\frac{2}{4}$

How did you decide?

____

____

5. How many faces does a square-based pyramid have?

________ faces

6. Write the number of faces, edges, and corners.

	faces	edges	corners
triangular prism			
cube			
triangular pyramid			

Name Date

Investigating Symmetry

1. For each pair of lines, draw 2 or more lines to make a shape that is symmetrical.

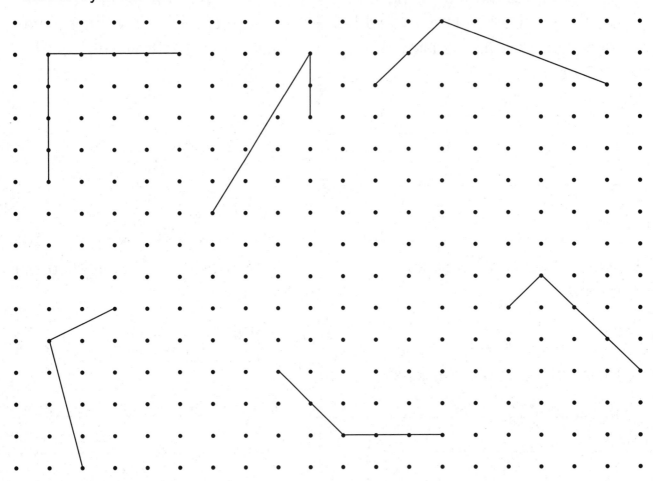

2. Place a ✓ by a shape you made that has more than 1 line of symmetry. Draw all the lines of symmetry on that shape.

3. Shade squares on the other side of the balance line to make each design symmetrical.

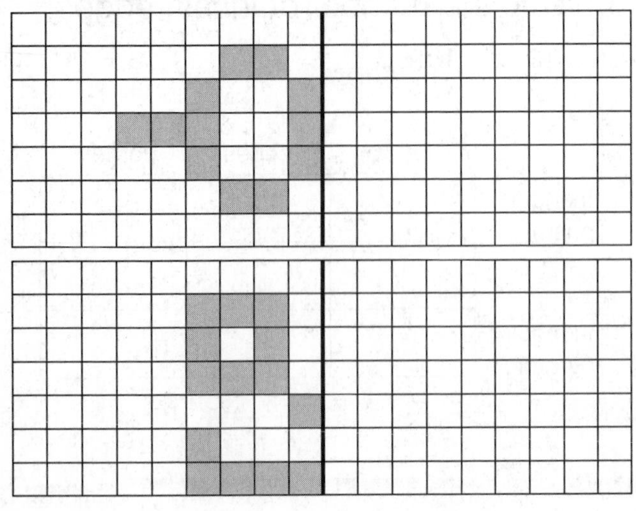

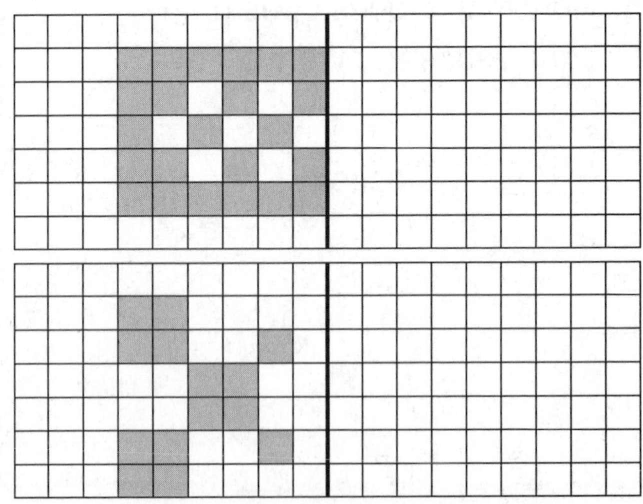

Name Date

Maintaining Concepts and Skills

1. Measure the lines in centimeters. Write the answers.

_______ cm

_______ cm

_______ cm

2. Use a centimeter ruler. Find the perimeter.

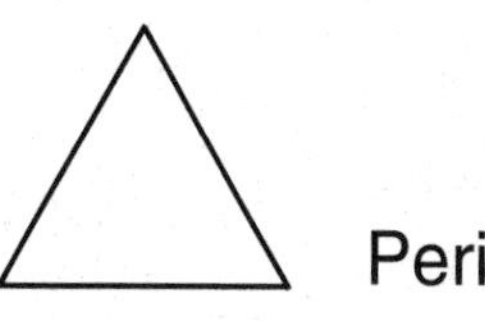

Perimeter _______ cm

3. Jasmine bought 6 roses and 6 carnations. How much did she spend?

$__________

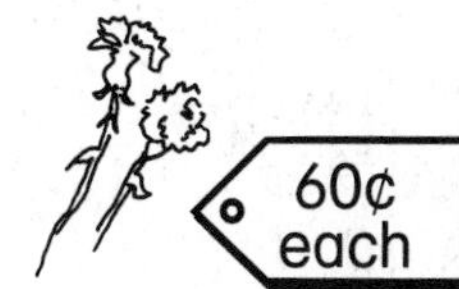

4. If you buy a pen for $1.65 (including tax) and give the salesperson $5.00, how much change will you get?

$__________

5. Ring pairs of numbers that you could multiply to make the target number. (Use a different color for each pair.)

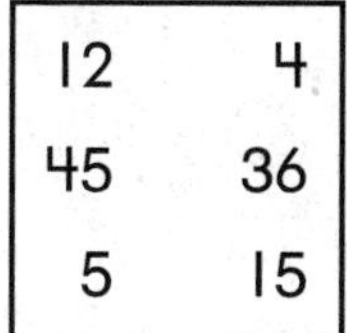

6. Write a letter on each shape.
- Cube – A
- Cone – B
- Triangular prism – C

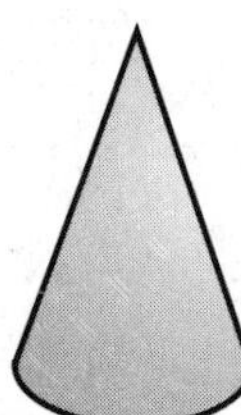

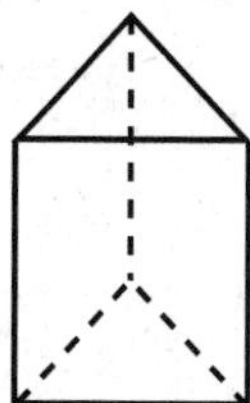

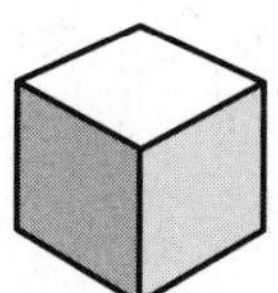

Name Date

Maintaining Concepts and Skills

1. Solve these.

4 × 26 = ______

6 × 32 = ______

41 × 5 = ______

19 × 7 = ______

2. Kerri saved quarters for a month. She had 80 quarters. How much money did she save?

$__________

3. Pretend you have 36 inches of string. How long would each side be if you made a

square? __________ in.

triangle? __________ in.

hexagon? __________ in.

4. Draw a rectangle. Then draw lines to show how you could cut it into a square and 2 triangles.

5. Ring the unit that would be best to measure the height of a tall building.

mile foot inch pound

6. Alta has 8 coins. The value is $1.15. What coins might she have?

____

____

____

Name Date

Using a Map

Some fourth-grade students are planning an orientation activity. This map shows the path they will follow.

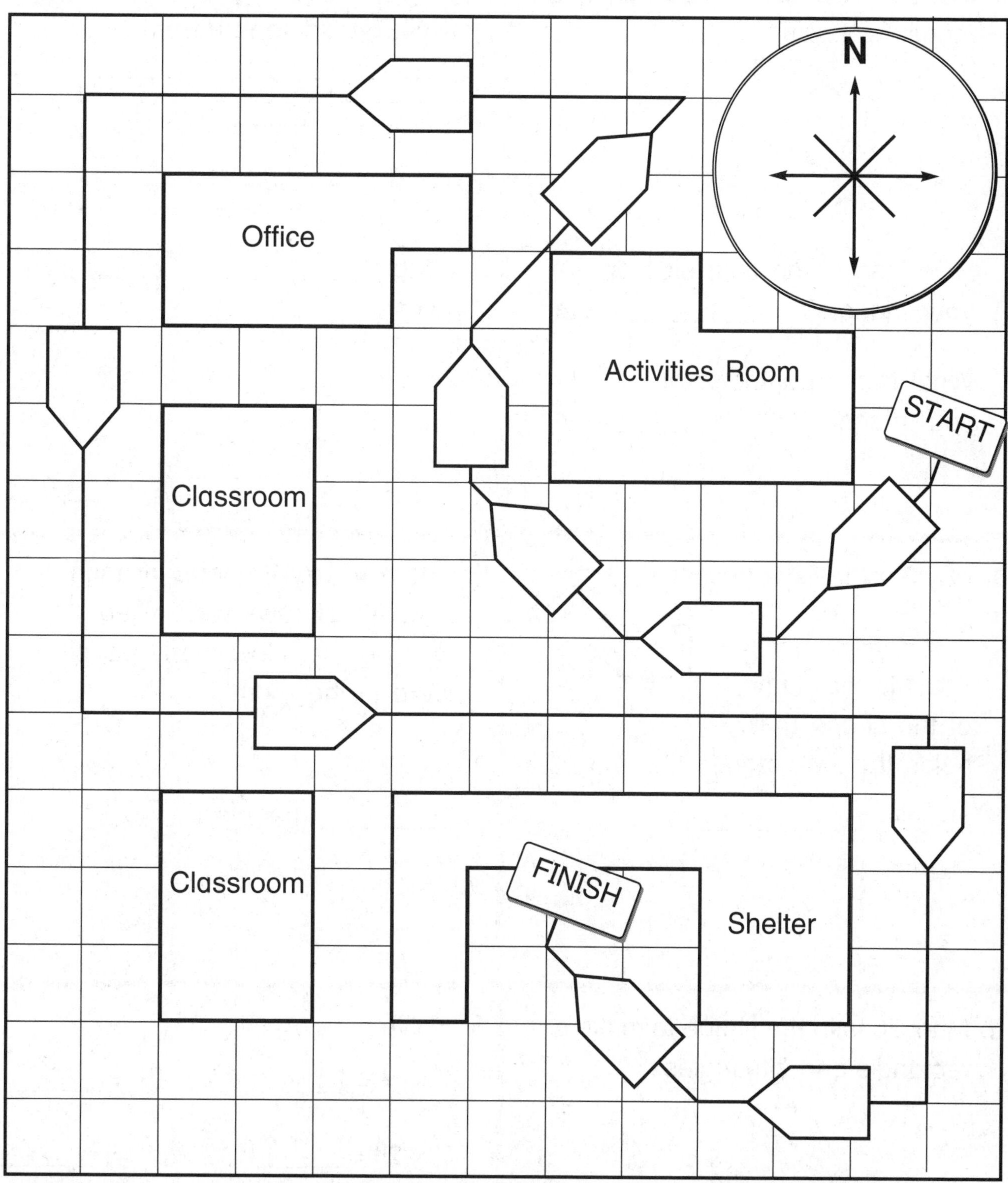

1. Write S., E., W., N.E., S.E., N.W., and S.W. on the compass.
2. Follow the paths from the start. Write on each arrow the direction it is pointing.

Name Date

Maintaining Concepts and Skills

1. Use a centimeter ruler. Draw a triangle that has 2 sides that each measure 4 cm.

How long is the third side on your triangle? ____________ cm

What is the perimeter of your triangle? ____________ cm

2. Estimate how many centimeters long this line segment would be if you straightened it out.

Length ________________ cm

3. Buy 3. What is your cost?

$________________

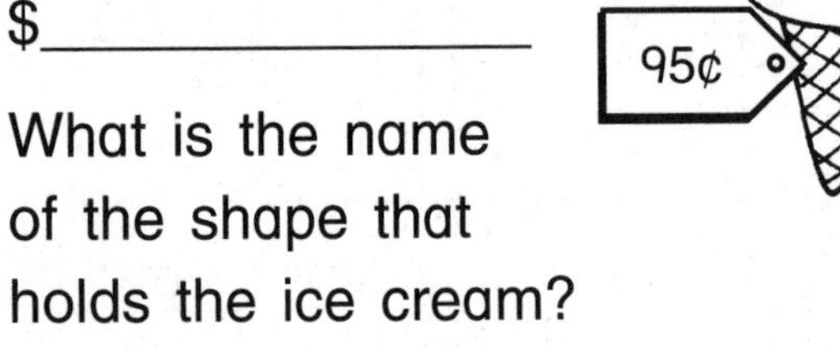

What is the name of the shape that holds the ice cream?

____

4. There were 12 chairs in each row. All 21 rows were filled with people. How many people were in the room?

_________ people

5. Mark lines in the shape to make a rectangle and 2 triangles.

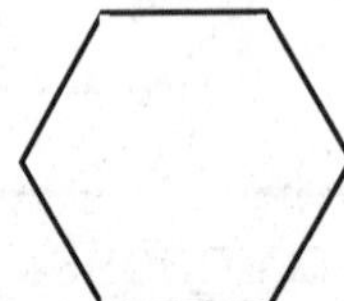

6. Solve.

7 × 4 = ______   7 × 14 = ______

7 × 24 = ______   7 × 34 = ______

Name Date

Creating and Investigating Triangles

1. Connect the 3 points in each circle to make a triangle.

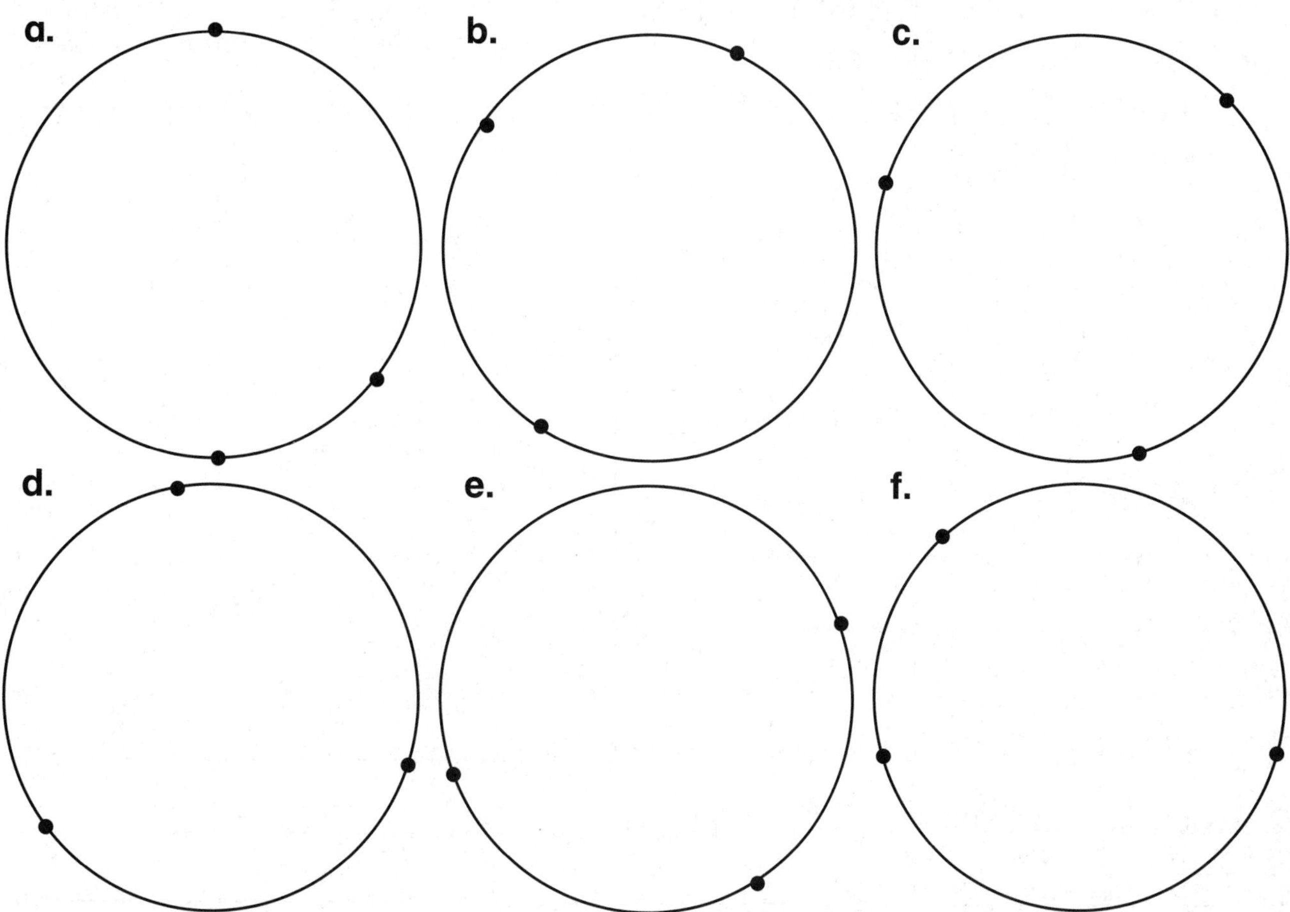

2. Find the 2 triangles that each have 3 equal angles.
   Color those triangles red.
3. Find the 2 triangles that each have 2 equal angles.
   Color those triangles blue.
4. Find the 2 triangles that each have a right angle.
   Color those triangles yellow.
5. Draw the following triangles on the dot paper below.
   a. a triangle that has one angle greater than a right angle
   b. a triangle with one right angle and 2 equal sides

a. **b.**

Name Date

Investigating and Creating Angles

1. Look at the angles in the shapes below.
   - Place an ✗ by all the acute angles.
   - Ring all the right angles.
   - Place a ✓ by all the angles that are greater than a right angle.

Acute Angle

An angle smaller than a right angle is called an acute angle.

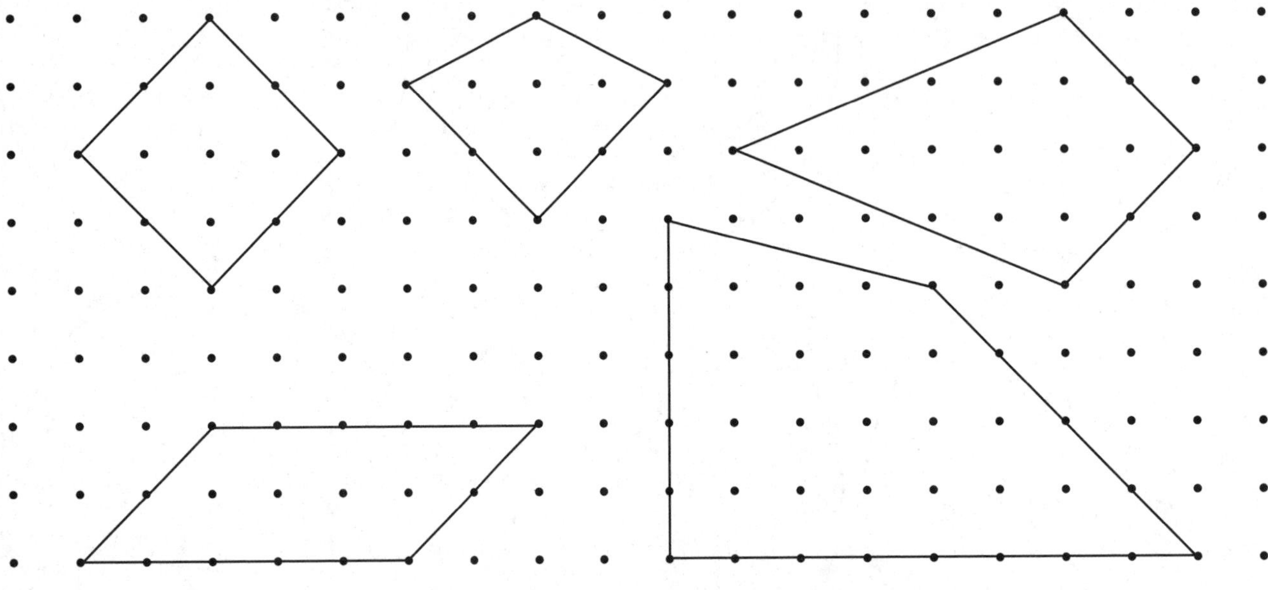

2. What do you think an obtuse angle is? ______________________________

____

3. Draw each of these shapes.
   a. a quadrilateral with exactly one right angle
   b. a quadrilateral with exactly 2 acute angles
   c. a quadrilateral with exactly one obtuse angle

a.

b.

c.

Name Date

Identifying Shapes

1. Read each shape "autobiography."
Label each shape with its correct name.

I am a 4-sided shape.
I have 2 pairs of parallel sides that are not the same length.
I am called a **parallelogram**.

I am a **trapezoid**.
I also have 4 sides.
I only have one pair of parallel sides.

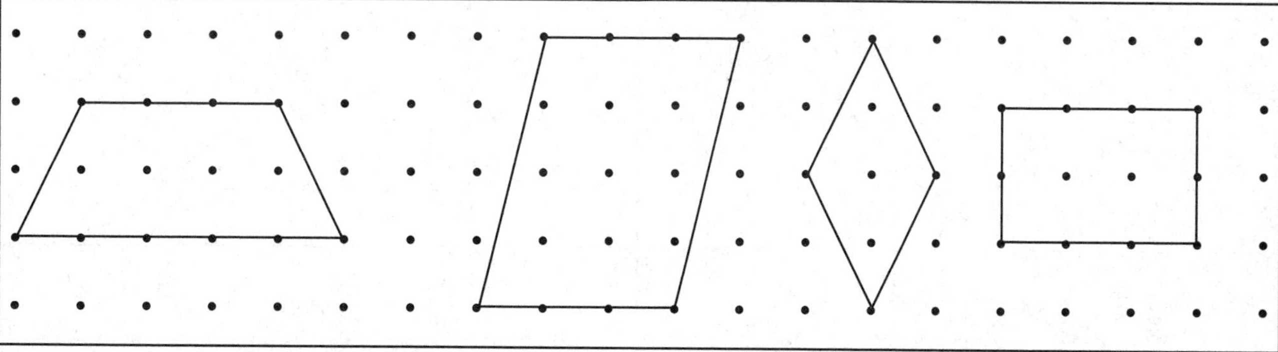

I am a special sort of parallelogram.
I have 2 pairs of parallel sides.
All my sides are the same length.
I am a **rhombus**.

I am a special sort of parallelogram.
I have 2 pairs of parallel sides.
I have 4 right angles.
I am a **rectangle**.

2. Draw a parallelogram, a trapezoid, a rhombus, and a rectangle.
Label each shape.

Name Date

Identifying Angles

For these shapes

- Draw a ● at each right angle.
- Place a ✔ by each acute angle.
- Place an ✘ by each obtuse angle.

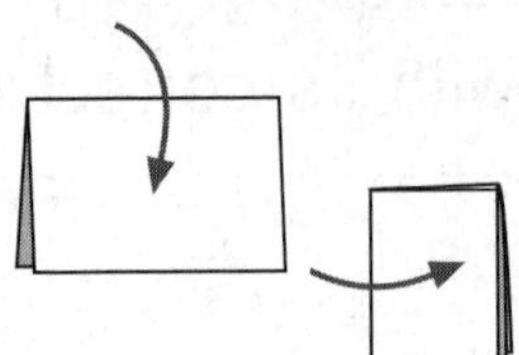

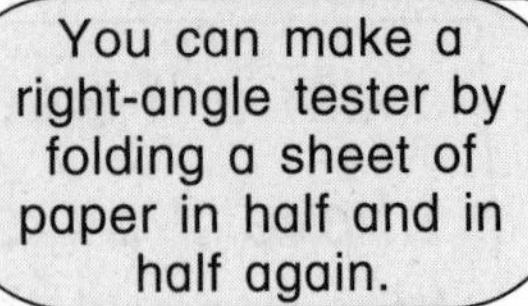

You can make a right-angle tester by folding a sheet of paper in half and in half again.

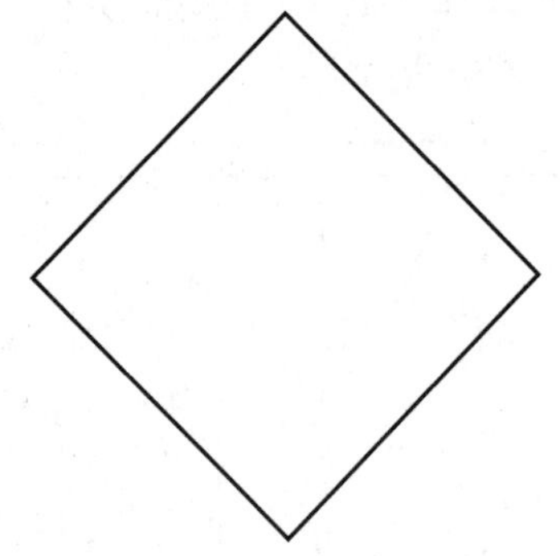

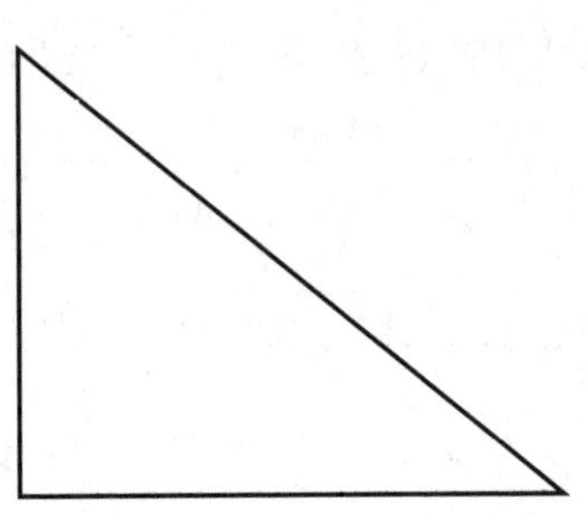

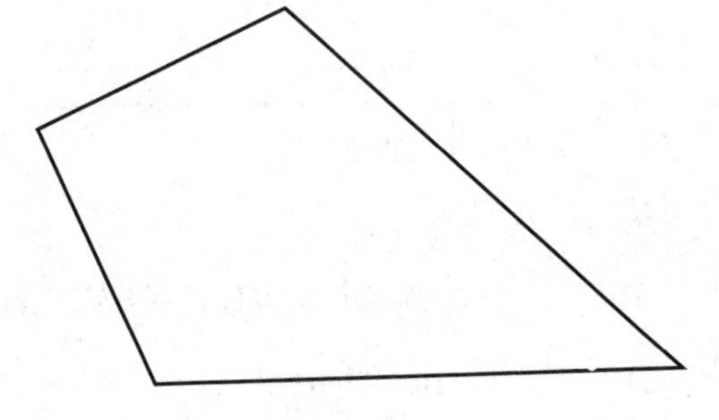

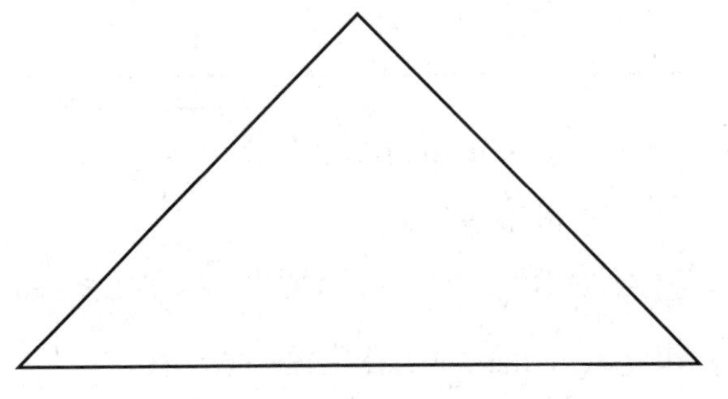

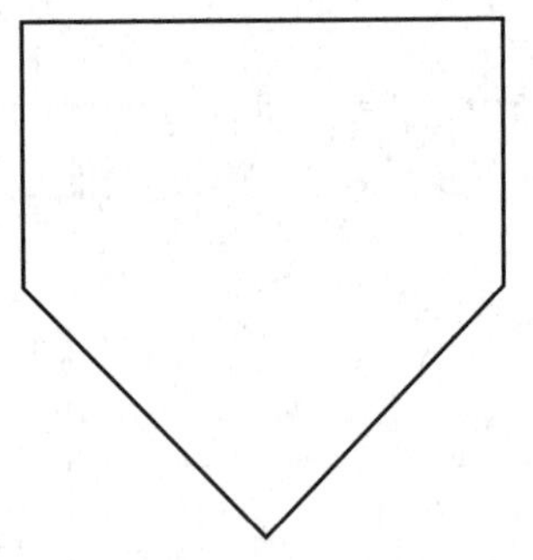

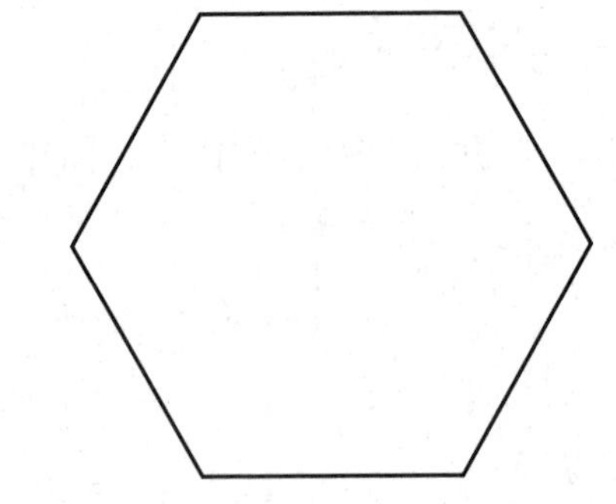

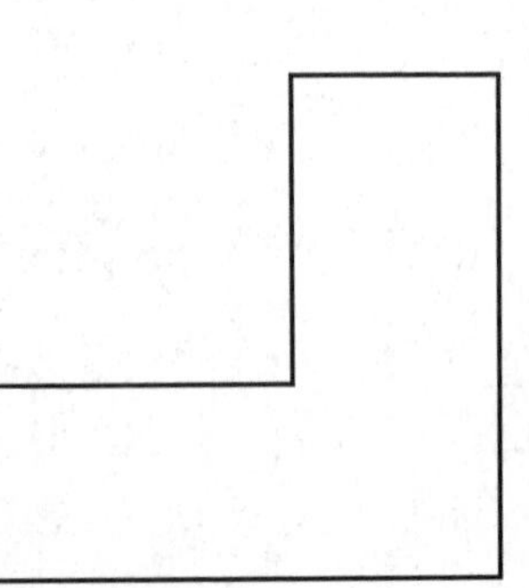

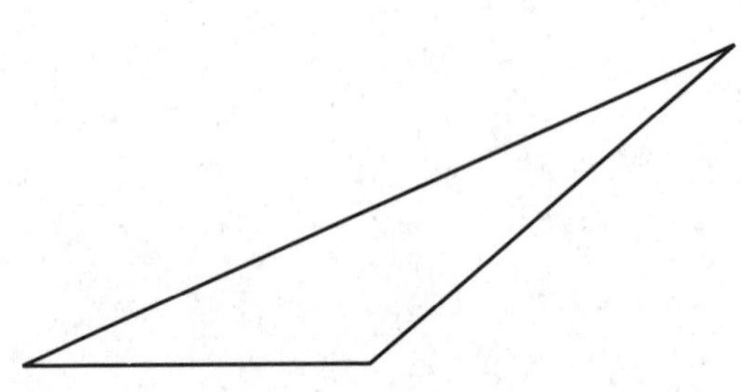

Use with Investigation 5.5

Name Date

Maintaining Concepts and Skills

1. The ecology club collected cans to recycle. They collected about 35 pounds of cans each week. About how many pounds did they collect in a month?

   _________ pounds

2. If the club gets paid 50 cents per pound for its cans, how much money does it get in one week?

   $_______________

   How much in a month? $_________

3. Duncan cut 3 boards. Each board was 4 feet, 2 inches long. How much wood did he use?

   ___________ ft ___________ in.

4. Estimate the width of the top of a classroom desk or table.

   ____

   What is something else that is about the same width?

   ____

5. Kara Kangaroo jumps 16 feet. How far has she jumped after

   a. 3 jumps? _______________

   b. 6 jumps? _______________

   c. 9 jumps? _______________

   d. 12 jumps? _______________

6. How can the answer to letter **a** in question 5 help you find the answers to **b**, **c**, and **d**?

   ____

   ____

   ____

Name Date

Using Fractions

1. Write or draw the answers to these fraction problems.

4 friends shared 5 sandwiches.

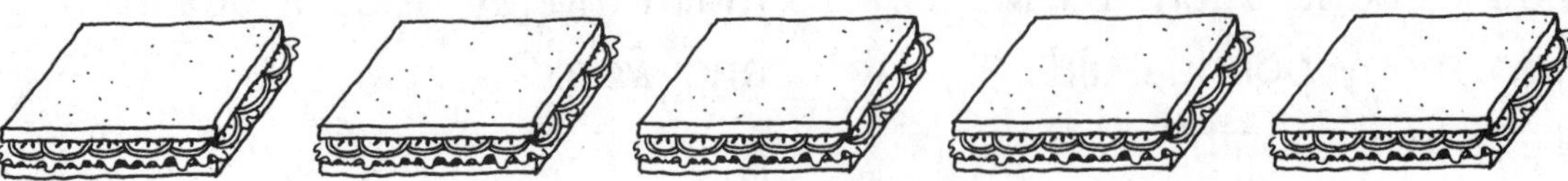

How many sandwiches did each person get? ______________________

6 students shared 3 bags of marbles.

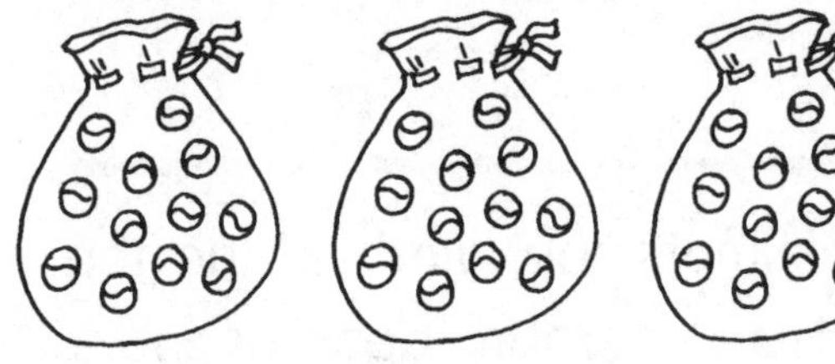

What **fraction of a bag** did each student get? ________________

How many marbles did each student get? ________________

3 friends shared 5 bananas.

How many bananas did each person get? ________________

2. Write a fraction story about this picture.

____

____

____

____

Name Date

Maintaining Concepts and Skills

1. Three children shared 39 pumpkin seeds. How many seeds did each child get?

_________ seeds

2. Estimate the answers.

52 ÷ 3 = _____

71 ÷ 5 = _____

49 ÷ 4 = _____

3. Use all of these digits to make each number.

9, 0, 8, 1, 5

the greatest number __________

a number with no tens ________

two even numbers

_____________ _______________

4. Select the best unit to measure

- the distance from New York City to Boston.

  yards miles feet

- the height of your school building.

  feet miles inches

5. Write these numbers in order from least to greatest.

789 10,001 9,009 9,999

____

6. Draw a rectangle with sides of 1 cm, 6 cm, 1 cm, and 6 cm.

What is its perimeter? ______________

Name Date

Finding a Fraction of an Amount

Use a different color to shade each person's share of pizza. Then write the amount each person got.

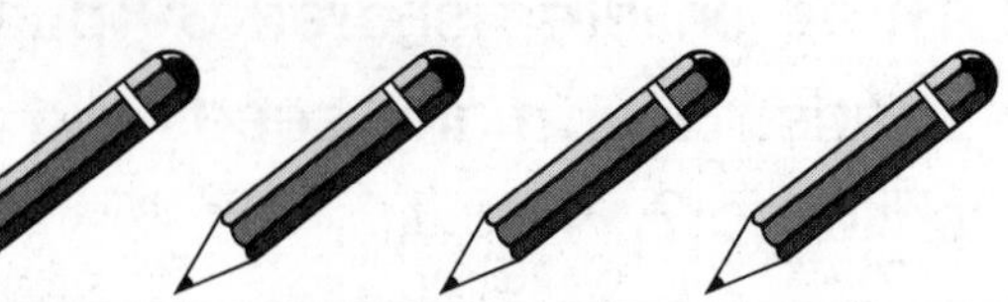

2 friends shared 5 pizzas. 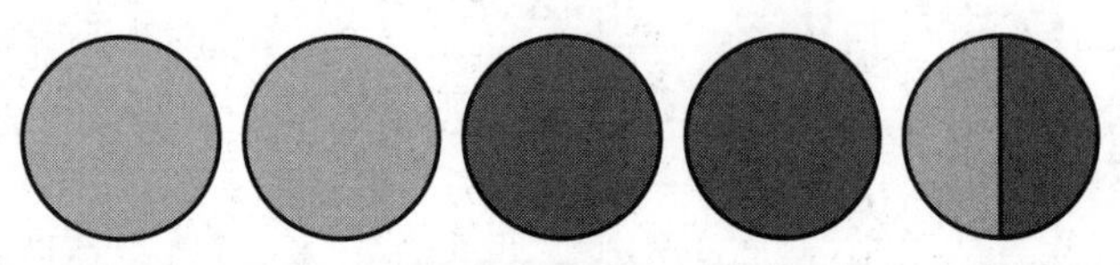	$\frac{1}{2}$ of 5 = $2\frac{1}{2}$
2 friends shared 3 pizzas. 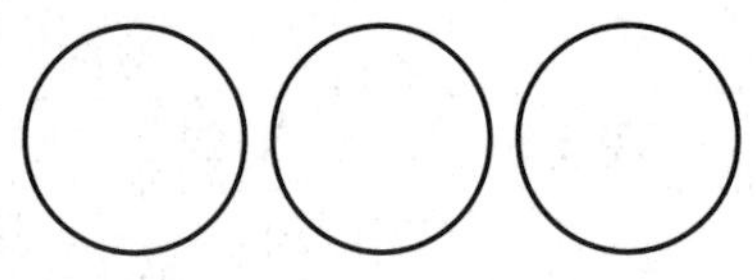	$\frac{1}{2}$ of 3 = _____
4 friends shared 5 pizzas. 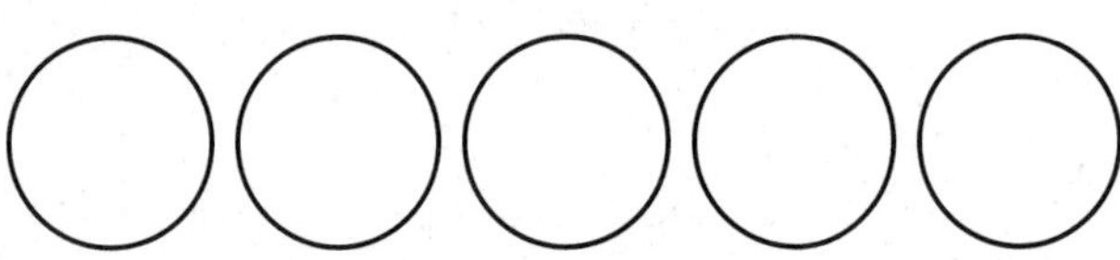	$\frac{1}{4}$ of 5 = _____
4 friends shared 7 pizzas. 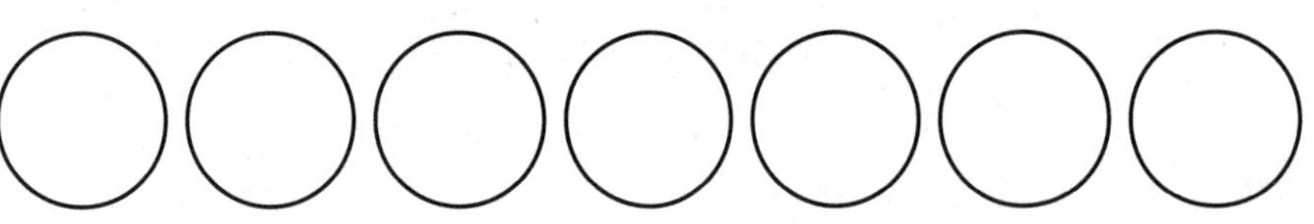	$\frac{1}{4}$ of 7 = _____
4 friends shared 3 pizzas. 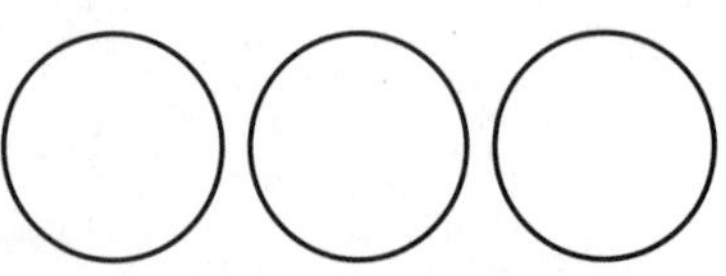	$\frac{1}{4}$ of 3 = _____

Name Date

Finding a Fraction of an Amount

Use a different color to shade each person's share of pizza. Then write the amount each person got.

3 friends shared 4 pizzas. 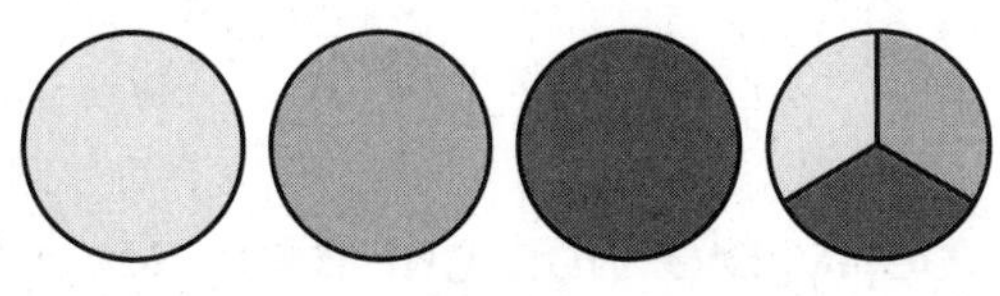	$\frac{1}{3}$ of 4 = $1\frac{1}{3}$
3 friends shared 7 pizzas. 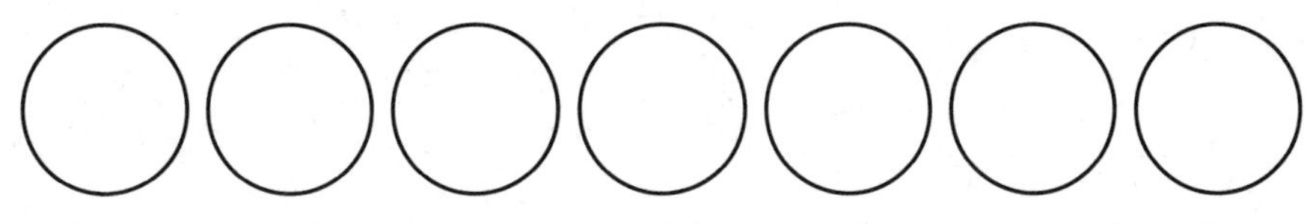	$\frac{1}{3}$ of 7 = _____
3 friends shared 5 pizzas. 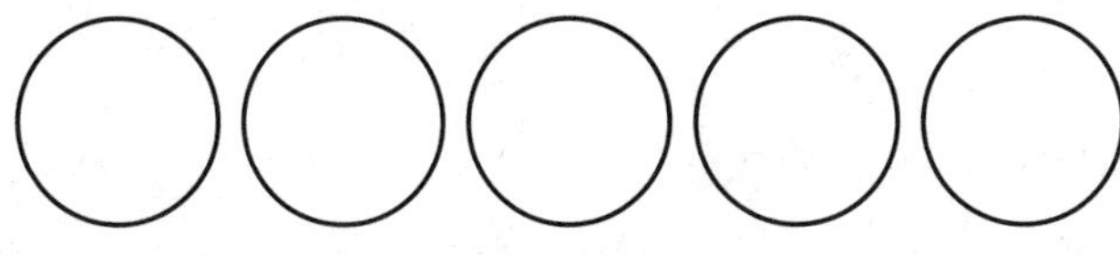	$\frac{1}{3}$ of 5 = _____
4 friends shared 6 pizzas. 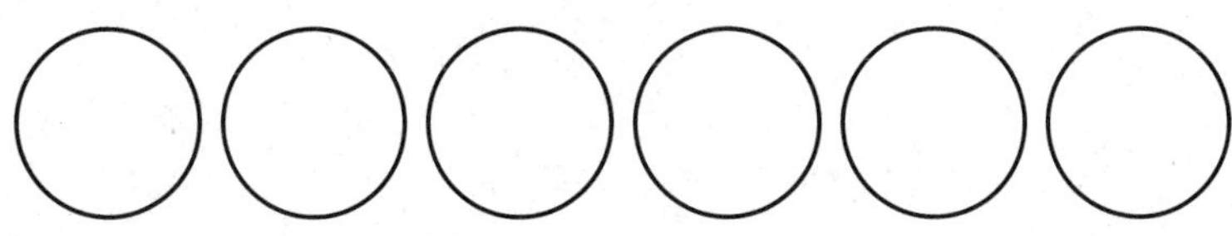	$\frac{1}{4}$ of 6 = _____
2 friends shared $3\frac{1}{2}$ pizzas. 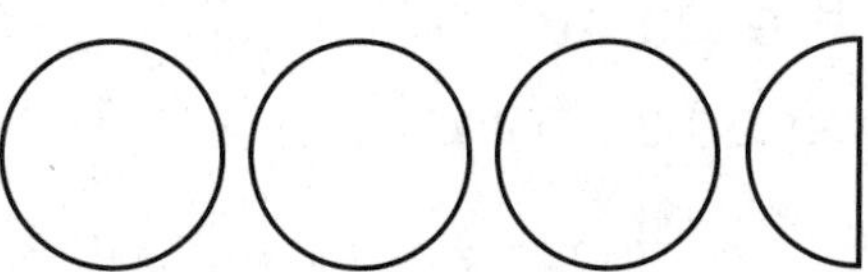	$\frac{1}{2}$ of $3\frac{1}{2}$ = _____

For this activity, students use a pictorial model to help them find a fraction of an amount. If your child needs to use a concrete model, he or she could cut out paper circles or squares and work with those.

Name Date

Solving Fraction Problems

Draw pictures to show how to solve these fraction problems.

1. After Tammy made her model, she had $3\frac{2}{3}$ sticks of modeling clay left over. Draw the clay that was left over.

2. There were 2 pies the same size. The strawberry pie was cut into 10 pieces. The apple pie was cut into 8 pieces.
   Derrick ate 3 of the 10 pieces of strawberry pie.
   Tanya ate 1 of the 8 pieces of apple pie.
   Who ate more pie, Derrick or Tanya? ______________________

3. It is 5 miles from Ellie's house to the movie theater.
   Jane's house is halfway between Ellie's house and the theater.

   How far is Jane's house from Ellie's? ______________

4. Tisha won $\frac{1}{4}$ of the special coins in a game.
   If she won 5 special coins, how many were in the game to start with? _______ special coins

Name Date

Maintaining Concepts and Skills

1. Use the numbers in the box to make division sentences that have answers between 5 and 10.

24	7	72	6	49
4	8	9	7	

____

____

2. Find the answers for these and write a story problem for one.

$72 \times 2 =$ _____ $87 \div 3 =$ _____

____

____

____

____

____

3. Estimate the length in inches. Then use a ruler to measure the line segments.

_____ estimate _____ inches

_____ estimate _____ inches

4. Draw a square with a perimeter of 12 units.

5. Find a number that fits the clues.

I am more than 1,000.
Three of my digits are 8, 2, 7.
I am less than 1,700.
I am an odd number. _______

6. Jerri put 48 cookies into bags to sell. There were 6 cookies in each bag. How many bags did she use?

_____ bags

Name Date

Creating Fractional Parts

Each [small dot grid] shows a school's play area. Draw lines so that the play area is shared equally between the number of classes shown. Then write the fraction of the playground each class is given.

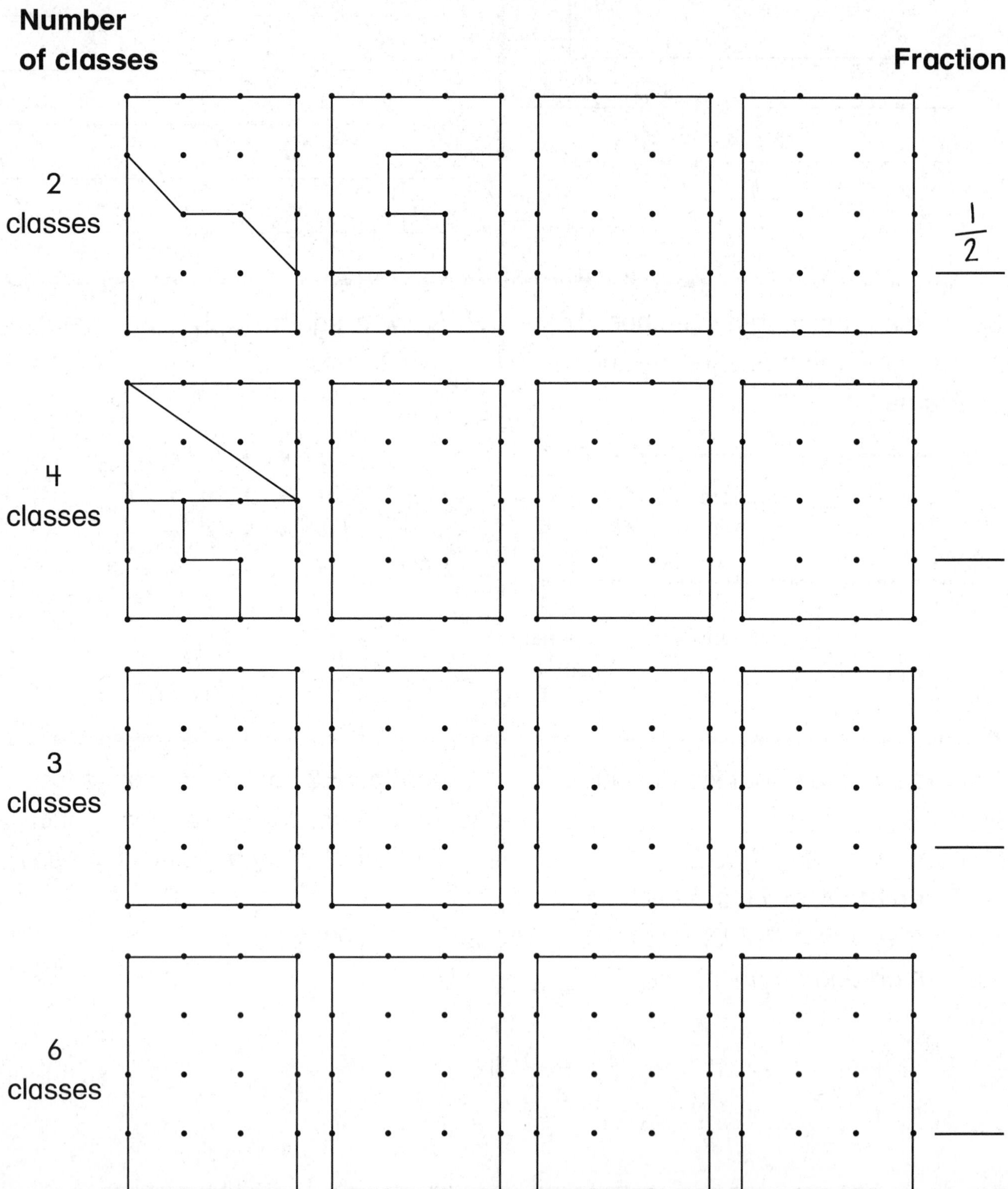

Name Date

Maintaining Concepts and Skills

1. Four children shared a long piece of licorice. Each child got 6 centimeters of licorice. How long was the piece?

____

2. A piece of rope 96 meters long is cut into 4 equal parts. How long is each part?

____

3. Estimate, using the unit you think is best.

How long is a bicycle? ________

How long is a passenger car?

____

Explain how you decided your estimates.

____

____

4. Use a calculator to fill the blanks.

36 ÷ 4 = ______

______ ÷ 5 = 9

84 ÷ ______ = ______

55 inches = ______ feet and

______ inches

5. Answer the questions below for these numbers.

98 198 98,001 1,098

9,890 10,098 981 10,908

Which is the greatest? ________

Which two have the least difference?

____

6. Put numbers in the blanks to make division sentences.

_____ ÷ _____ = _____

_____ ÷ 3 = 12

63 ÷ _____ = _____

_____ ÷ 7 = 7

Name Date

Finding Equivalent Fractions

Write the missing number to make equivalent fractions.
Use the fraction strips to help.

Fraction strips

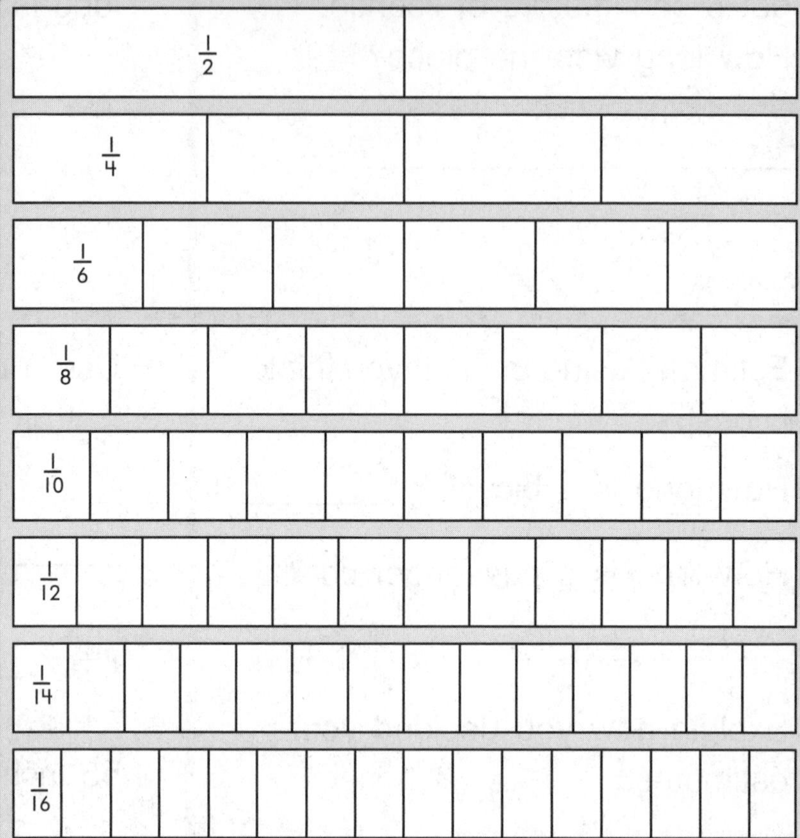

$\frac{1}{2} = \frac{3}{6}$

$\frac{1}{2} = \frac{}{4}$

$\frac{3}{4} = \frac{}{8}$

$\frac{3}{8} = \frac{}{16}$

$\frac{1}{6} = \frac{}{12}$

$\frac{3}{4} = \frac{}{12}$

$\frac{1}{4} = \frac{}{8} = \frac{}{12} = \frac{}{16}$

$\frac{1}{2} = \frac{}{16} = \frac{}{14} = \frac{}{12} = \frac{}{10} = \frac{}{8}$

$\frac{14}{16} = \frac{}{8}$

$\frac{12}{16} = \frac{}{4}$

$\frac{4}{16} = \frac{}{8}$

$\frac{8}{12} = \frac{}{6}$

$\frac{10}{12} = \frac{}{6}$

$\frac{6}{8} = \frac{}{4}$

Name Date

Finding Equivalent Fractions

Write the missing number to make equivalent fractions.
Use the fraction strips to help.

Fraction strips

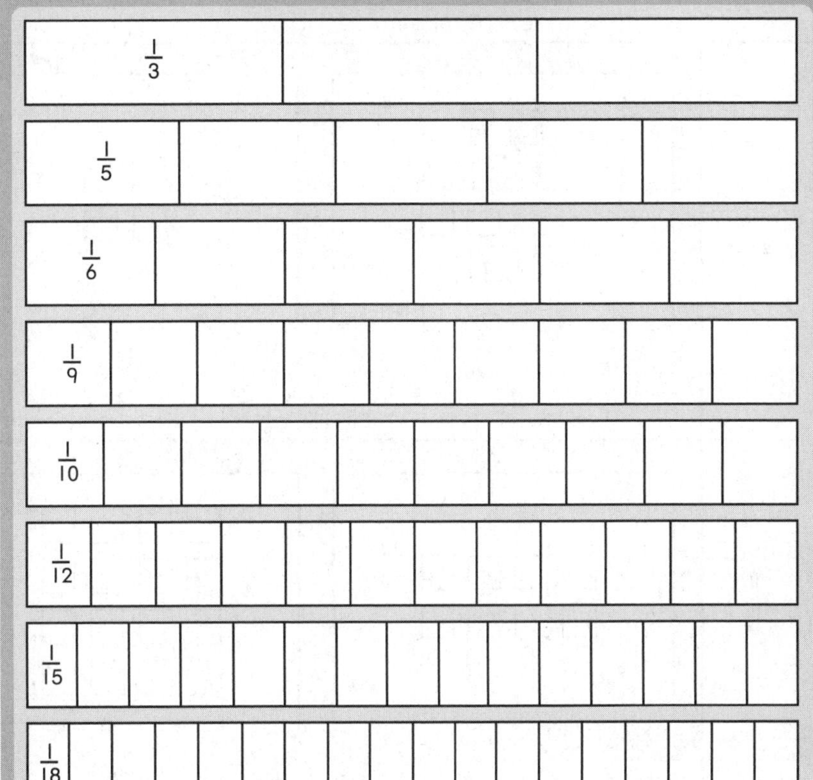

$\frac{2}{3} = \frac{\quad}{6}$

$\frac{2}{3} = \frac{\quad}{9}$

$\frac{2}{5} = \frac{\quad}{10}$

$\frac{3}{5} = \frac{\quad}{15}$

$\frac{5}{6} = \frac{\quad}{12}$

$\frac{1}{9} = \frac{\quad}{18}$

$\frac{1}{5} = \frac{\quad}{10} = \frac{\quad}{15}$

$\frac{1}{3} = \frac{\quad}{18} = \frac{\quad}{15} = \frac{\quad}{12} = \frac{\quad}{9} = \frac{\quad}{6}$

$\frac{16}{18} = \frac{\quad}{9}$

$\frac{15}{18} = \frac{\quad}{6}$

$\frac{12}{18} = \frac{\quad}{9}$

$\frac{10}{18} = \frac{\quad}{9}$

$\frac{9}{18} = \frac{\quad}{6}$

$\frac{8}{18} = \frac{\quad}{9}$

To compare, add, or subtract fractions that have different denominators, the students need to be able to find equivalent fractions. After completing the page, your child could use the fraction strips to find 5 other pairs of equivalent fractions.

Name Date

Creating Equivalent Fractions

Look at each fraction. Draw and shade cubes in the 3 rectangles to show that fraction in **3** different ways. Write each fraction you showed.

$\frac{1}{2}$	$\frac{6}{12}$	$\frac{2}{4}$	______
$\frac{1}{4}$	______	______	______
$\frac{1}{3}$	______	______	______
$\frac{3}{4}$	______	______	______

Name Date

Maintaining Concepts and Skills

1. Find two numbers that will divide each of these numbers evenly (with no remainders).

   32 52 44 40 48 36

   _____ _____

   Tell how you found the numbers.

   ____

   ____

   ____

   ____

2. The grocery store sold 125 dozen eggs to 5 people. Each person got the same amount. How many dozens did each get?

   ________ dozens

3. What number is 1,000 more than 305?

   ____

4. Write three numbers greater than 10,000 and one thousand less than each other.

   ________ ________ ________

5. Estimate or use mental arithmetic to change the units.

   _____ inches = 4 feet

   72 inches = _____ yards

   _____ inches = 3 yards

   8 yards = _____ feet

6. Estimate the length in centimeters. Then use a ruler to measure the line segments.

   ____

   _____ estimate _____ centimeters

   ____

   _____ estimate _____ centimeters

Name Date

Creating Multiplication Arrays

1. For each of these:
- Shade the grid to show the first fact.
- Write the answer for both facts.

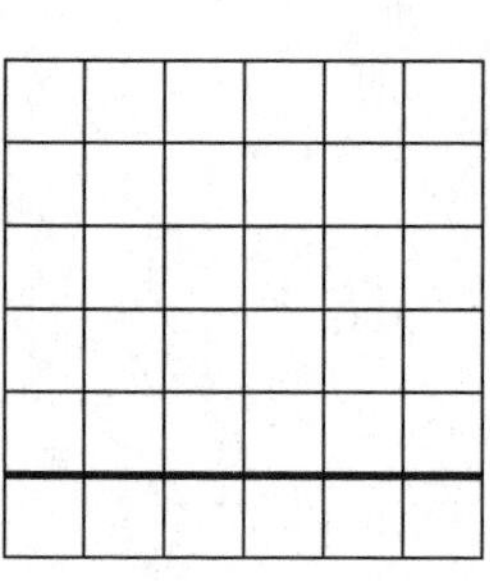

5 × 6 = ____

6 × 6 = ____

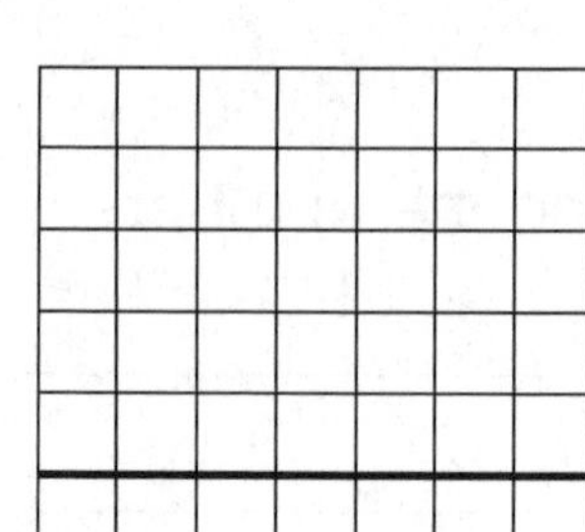

5 × 7 = ____

6 × 7 = ____

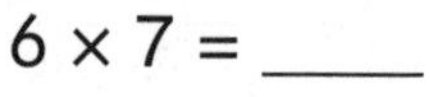

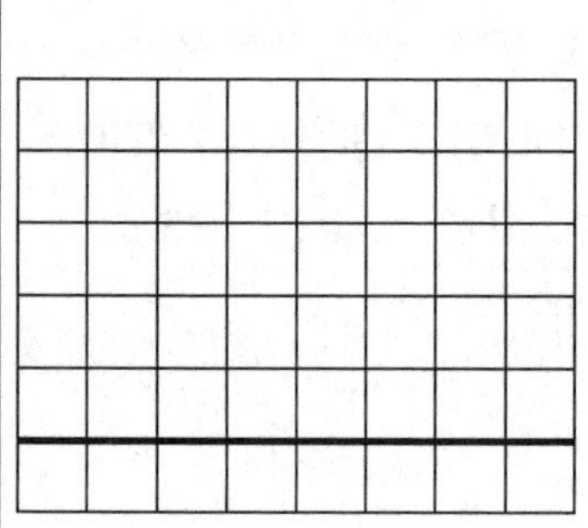

5 × 8 = ____

6 × 8 = ____

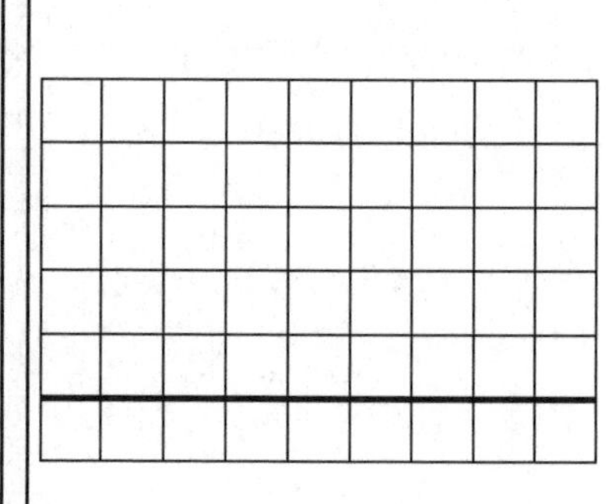

5 × 9 = ____

6 × 9 = ____

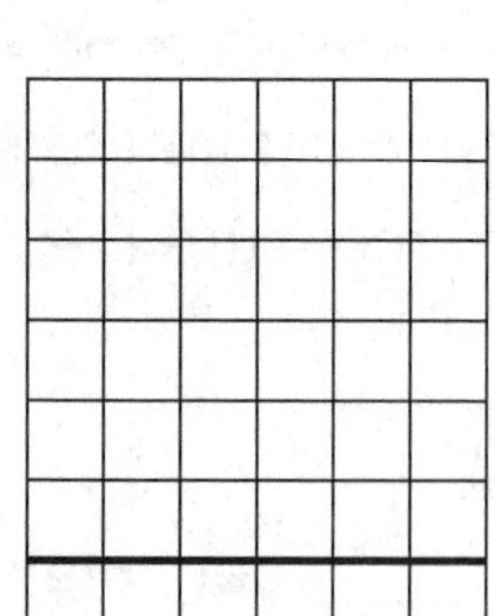

6 × 6 = ____

7 × 6 = ____

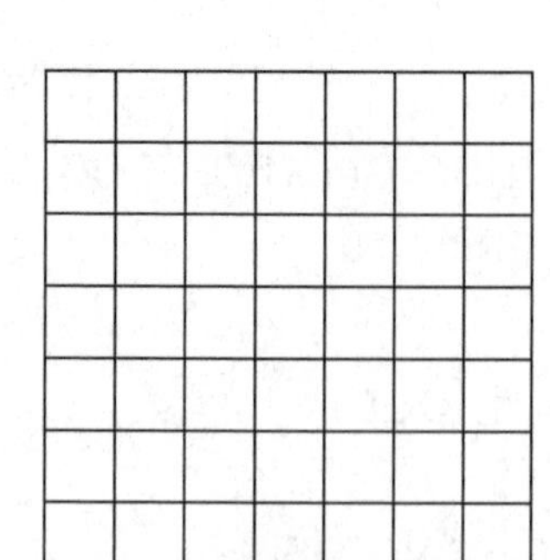

7 × 7 = ____

8 × 7 = ____

2. Write how you figured out the answer to **8 × 7**.

____

____

Name Date

Maintaining Concepts and Skills

1. Write your answers for each shape. Use a centimeter ruler to measure.

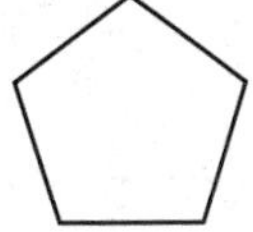

Number of sides ____
Number of angles ____
Perimeter ______ cm
Name ____________

Number of sides ____
Number of angles ____
Perimeter ______ cm
Name ____________

2. Choose the best unit of measure for each item from the box below. Use each unit only once.

weight of a candy bar ______
your height ______
distance across an ocean ______
your weight ______
capacity of a milk carton ______
distance a ball is thrown ______

centimeter	liter	kilogram
gram	meter	kilometer

3. Use these digits to make the greatest number you can.

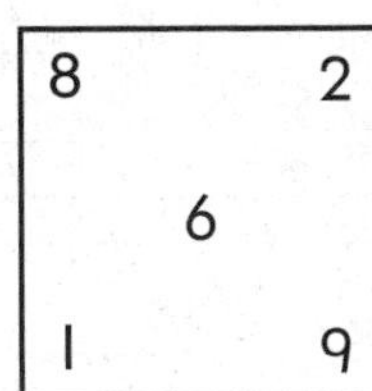

____

4. a. What is the greatest number you can make using 5 different digits?

____

b. What is the smallest number you can make using 5 different digits except 0?

____

Name Date

Practicing Multiplication Number Facts

Write the answers. (You can use the answer to the first fact to help figure out the second fact.)

$2 \times 4 =$ _____	$3 \times 4 =$ _____	$2 \times 5 =$ _____	$3 \times 5 =$ _____
$5 \times 3 =$ _____	$6 \times 3 =$ _____	$10 \times 2 =$ _____	$9 \times 2 =$ _____
$2 \times 6 =$ _____	$3 \times 6 =$ _____	$5 \times 7 =$ _____	$6 \times 7 =$ _____
$10 \times 4 =$ _____	$9 \times 4 =$ _____	$2 \times 8 =$ _____	$3 \times 8 =$ _____
$5 \times 8 =$ _____	$6 \times 8 =$ _____	$10 \times 6 =$ _____	$9 \times 6 =$ _____
$9 \times 2 =$ _____	$10 \times 2 =$ _____	$7 \times 2 =$ _____	$8 \times 2 =$ _____
$10 \times 8 =$ _____	$9 \times 8 =$ _____	$1 \times 2 =$ _____	$0 \times 2 =$ _____
$5 \times 2 =$ _____	$6 \times 2 =$ _____	$10 \times 3 =$ _____	$9 \times 3 =$ _____
$1 \times 7 =$ _____	$0 \times 7 =$ _____	$8 \times 2 =$ _____	$9 \times 2 =$ _____

Name Date

Practicing Multiplication Number Facts

Write the answers. (You can use the answer to the first fact to help figure out the second fact.)

$3 \times 2 =$ ______	$4 \times 2 =$ ______	$2 \times 3 =$ ______	$3 \times 3 =$ ______
$5 \times 4 =$ ______	$6 \times 4 =$ ______	$5 \times 10 =$ ______	$6 \times 10 =$ ______
$5 \times 2 =$ ______	$6 \times 2 =$ ______	$4 \times 2 =$ ______	$5 \times 2 =$ ______
$1 \times 5 =$ ______	$0 \times 5 =$ ______	$10 \times 2 =$ ______	$9 \times 2 =$ ______
$2 \times 7 =$ ______	$3 \times 7 =$ ______	$1 \times 4 =$ ______	$0 \times 4 =$ ______
$10 \times 5 =$ ______	$9 \times 5 =$ ______	$5 \times 5 =$ ______	$6 \times 5 =$ ______
$6 \times 2 =$ ______	$7 \times 2 =$ ______	$2 \times 9 =$ ______	$3 \times 9 =$ ______
$5 \times 6 =$ ______	$6 \times 6 =$ ______	$10 \times 9 =$ ______	$9 \times 9 =$ ______
$1 \times 10 =$ ______	$0 \times 10 =$ ______	$5 \times 9 =$ ______	$6 \times 9 =$ ______

The students will soon begin to multiply larger numbers. To use a multiplication algorithm, they need to be able to answer number facts quickly and accurately. Ask your child several multiplication facts during spare moments each day—while traveling in the car, preparing meals, and so on.

Name Date

Identifying Numbers

1. Use the clues to find the mystery numbers in this hundred chart.

1	2	3	4	5	6	7	8	9	10
11	12	13	14	15	16	17	18	19	20
21	22	23	24	25	26	27	28	29	30
31	32	33	34	35	36	37	38	39	40
41	42	43	44	45	46	47	48	49	50
51	52	53	54	55	56	57	58	59	60
61	62	63	64	65	66	67	68	69	70
71	72	73	74	75	76	77	78	79	80
81	82	83	84	85	86	87	88	89	90
91	92	93	94	95	96	97	98	99	100

- *I am less than 70 but greater than 50.*
- Both my digits are odd.
- *I am a multiple of 11.*
- Place an ✗ on me.

- *I am an even number.*
- *I am a square number.*
- The sum of my digits is 10.
- Color me blue.

- *I am an even number less than 20.*
- *I am a multiple of 2, 3, and 4.*
- My digits are in counting order.
- Color me yellow.

- *I am between 20 and 30.*
- *I am a prime number.*
- The difference between my digits is 7.
- Place a ✓ on me.

- *I am a multiple of 5 and 10.*
- *I have 12 factors.*
- The sum of my digits is 6.
- Ring me.

- The sum of my digits is 1.
- *I am a factor of 50.*
- *I am an even number.*
- Color me red.

2. Choose a number. Write your own clues and challenge a friend.

____

____

____

Name Date

Maintaining Concepts and Skills

1. Draw a rectangle with sides of 3 cm, 4 cm, 3 cm, and 4 cm.

   What is its perimeter? ______ cm

2. Use an inch ruler. Measure the length and width of your math book.

   length ________ in.

   width ________ in.

   What is the difference? ______ in.

3. Write the numbers.

   ten thousand, four hundred six

   ____

   two thousand, one hundred ninety-six

   ____

   thirty-four thousand, sixty-three

   ____

4. Put the numbers in order from least to greatest.

   1,996 989 2,001 1,001

   ______ ______ ______ ______

Name Date

Writing and Identifying Factors and Multiples

1. Write all the **factors** of each number.

8 ____ ____ ____ ____

18 ____ ____ ____ ____ ____ ____

28 ____ ____ ____ ____ ____ ____

35 ____ ____ ____ ____

2. Write 6 **multiples** of each number.

3 ____ ____ ____ ____ ____ ____

5 ____ ____ ____ ____ ____ ____

9 ____ ____ ____ ____ ____ ____

10 ____ ____ ____ ____ ____ ____

3. Place a ✔ by each **factor** of 16. Ring each **multiple** of 16.

1	2	3	4	5	6	7	8	9	10
12	16	20	24	28	32	36	40	44	48

Your child should be familiar with the words *factor* (for example, 2 is a factor of 6) and *multiple* (for example, 24 is a multiple of 6). Working with factors and multiples encourages students to explore the links between multiplication and division. These connections help students learn and remember number facts.

Name Date

Using Number Facts to Multiply Tens

Write the answers.

4×2	40×2	3×5	30×5	6×2	60×2
5×4	50×4	4×3	40×3	8×2	80×2
6×5	60×5	9×2	90×2	8×4	80×4
7×2	70×2	8×5	80×5	8×3	80×3
5×6	50×6	7×3	70×3	9×5	90×5

Name Date

Maintaining Concepts and Skills

1. Draw lines in each shape to cut it into triangles. (Lines should not cross each other.)

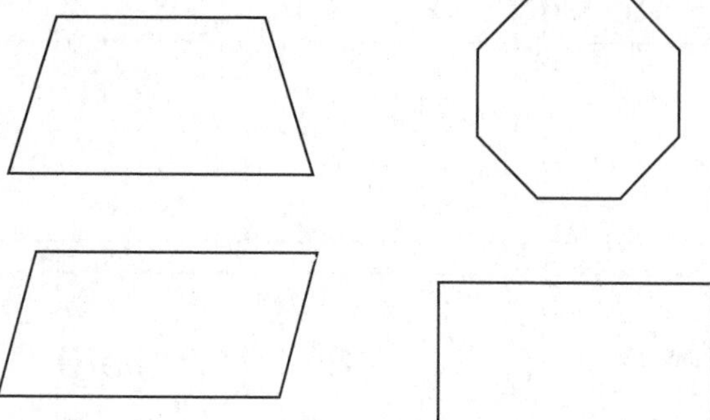

2. Alonte wanted to put reflecting tape on all of the edges of a large cube. One edge of the cube is 8 cm long. How much reflecting tape will he need?

____

Write how you found your answer.

____

____

____

3. Write the words for this number.

21,102

____

____

4. In the space below, draw 2 different shapes that have a perimeter of 12 cm.

Name Date

Using Number Facts to Multiply Tens

Write the answers.

$$\begin{array}{r} 3 \\ \times\ 2 \\ \hline \end{array} \qquad \begin{array}{r} 30 \\ \times\ 2 \\ \hline \end{array} \qquad \begin{array}{r} 4 \\ \times\ 5 \\ \hline \end{array} \qquad \begin{array}{r} 40 \\ \times\ 5 \\ \hline \end{array} \qquad \begin{array}{r} 6 \\ \times\ 3 \\ \hline \end{array} \qquad \begin{array}{r} 60 \\ \times\ 3 \\ \hline \end{array}$$

$$\begin{array}{r} 5 \\ \times\ 2 \\ \hline \end{array} \qquad \begin{array}{r} 50 \\ \times\ 2 \\ \hline \end{array} \qquad \begin{array}{r} 5 \\ \times\ 3 \\ \hline \end{array} \qquad \begin{array}{r} 50 \\ \times\ 3 \\ \hline \end{array} \qquad \begin{array}{r} 2 \\ \times\ 4 \\ \hline \end{array} \qquad \begin{array}{r} 20 \\ \times\ 4 \\ \hline \end{array}$$

$$\begin{array}{r} 7 \\ \times\ 5 \\ \hline \end{array} \qquad \begin{array}{r} 70 \\ \times\ 5 \\ \hline \end{array} \qquad \begin{array}{r} 2 \\ \times\ 6 \\ \hline \end{array} \qquad \begin{array}{r} 20 \\ \times\ 6 \\ \hline \end{array} \qquad \begin{array}{r} 2 \\ \times\ 5 \\ \hline \end{array} \qquad \begin{array}{r} 20 \\ \times\ 5 \\ \hline \end{array}$$

$$\begin{array}{r} 3 \\ \times\ 4 \\ \hline \end{array} \qquad \begin{array}{r} 30 \\ \times\ 4 \\ \hline \end{array} \qquad \begin{array}{r} 9 \\ \times\ 3 \\ \hline \end{array} \qquad \begin{array}{r} 90 \\ \times\ 3 \\ \hline \end{array} \qquad \begin{array}{r} 2 \\ \times\ 7 \\ \hline \end{array} \qquad \begin{array}{r} 20 \\ \times\ 7 \\ \hline \end{array}$$

$$\begin{array}{r} 5 \\ \times\ 5 \\ \hline \end{array} \qquad \begin{array}{r} 50 \\ \times\ 5 \\ \hline \end{array} \qquad \begin{array}{r} 3 \\ \times\ 6 \\ \hline \end{array} \qquad \begin{array}{r} 30 \\ \times\ 6 \\ \hline \end{array} \qquad \begin{array}{r} 4 \\ \times\ 4 \\ \hline \end{array} \qquad \begin{array}{r} 40 \\ \times\ 4 \\ \hline \end{array}$$

The examples on this page encourage the students to use number facts to help them multiply by "tens" (20, 30, 40, and so on). This strategy will be helpful when they explore techniques for multiplying other 2-digit numbers.

Name Date

Maintaining Concepts and Skills

1. The Middletown police estimated that the crowd for this year's July 4th fireworks was twenty-three thousand. Last year it rained before the fireworks, and the crowd was nearly nine thousand less. About how large was last year's crowd?

____

2. 9 runners ran a marathon race. The distance for a marathon is 26.2 miles, which is close to 25 miles. Make an estimate of the total miles run by all 9 runners.

__________________ miles

3. a. The distance around the earth at the equator is 24,903 miles. Write the number in words.

____

____

____

b. The distance across the United States is two thousand, five hundred seventy miles. Write the number.

____

4. Suppose you have 2 triangles and a square that have sides the same length. If you put them together along matching sides, how many sides will the new shape have?

________ sides

Draw it in the space below to test your answer.

Name Date

Using Arrays to Find Products

Look at the array in part **a.** Use it to help you find the product for 3 by 25. Follow the same steps to solve the other multiplication problems.

a.

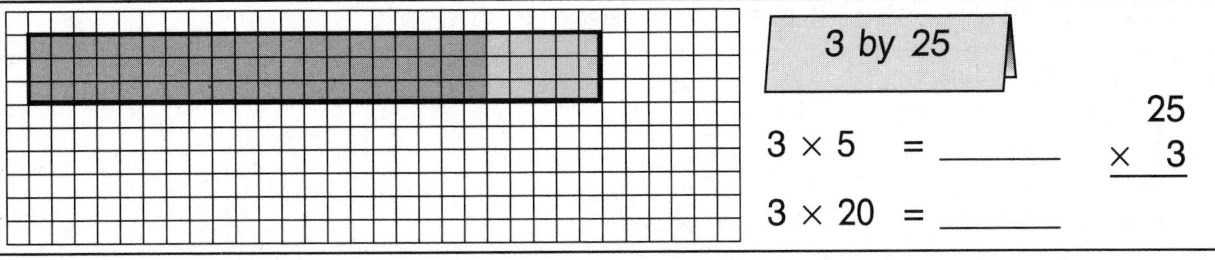

b.

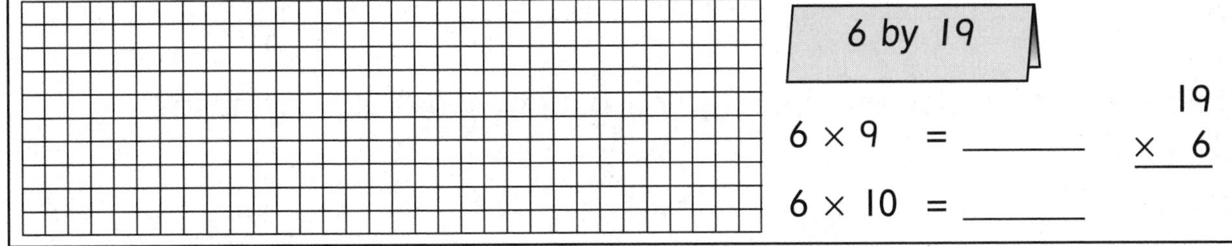

c.

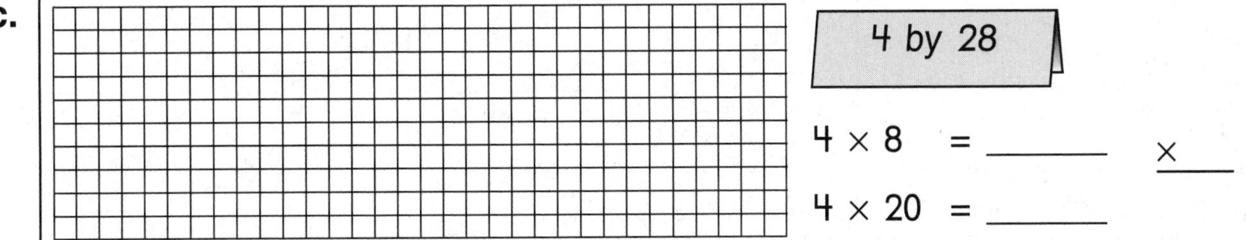

d.

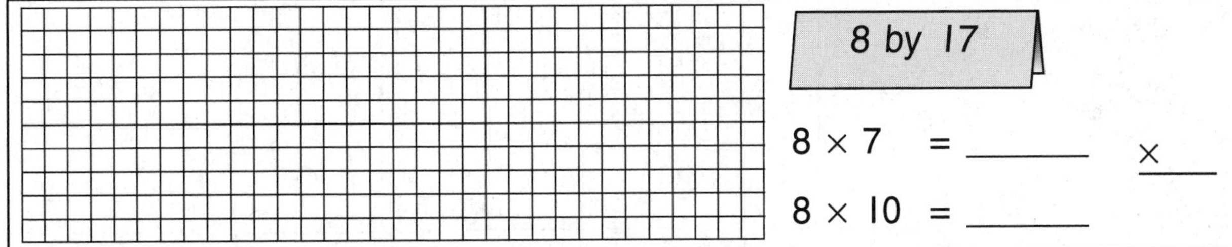

e.

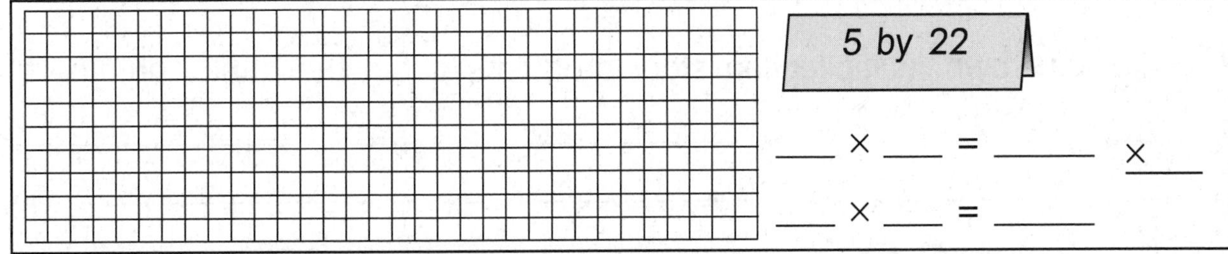

f.

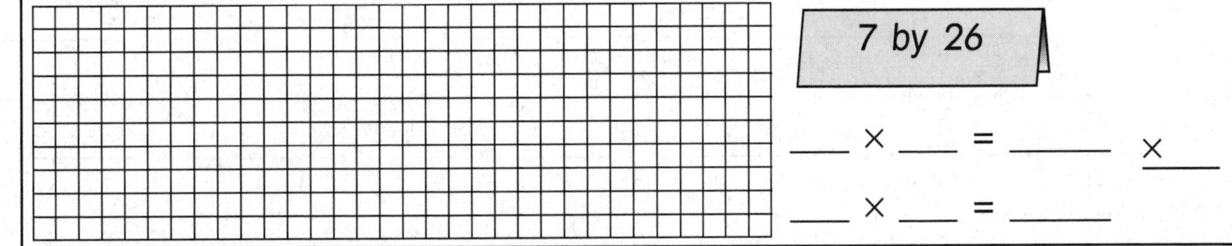

Name Date

Multiplying Costs

Cold Drinks

1. Calculate the cost of buying 3 of each of these.

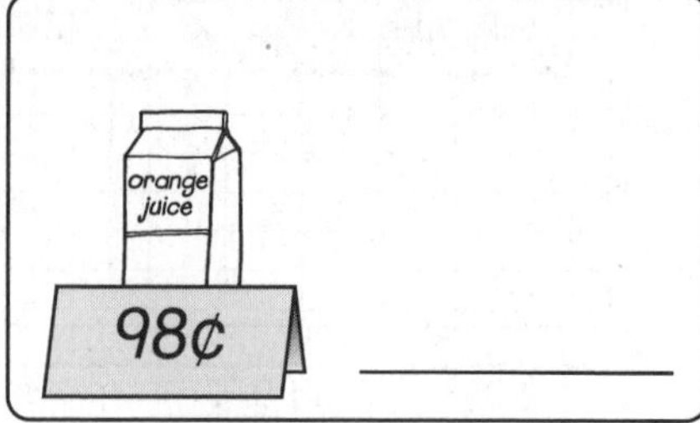

2. How much would you spend if you bought 6 of each of these?

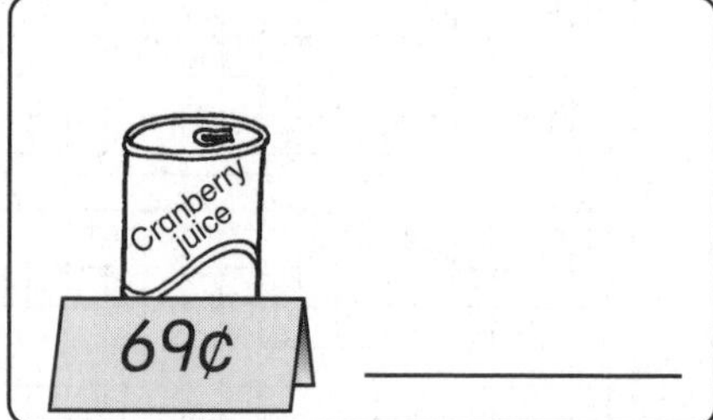

3. How much will 5 of each of these cost?

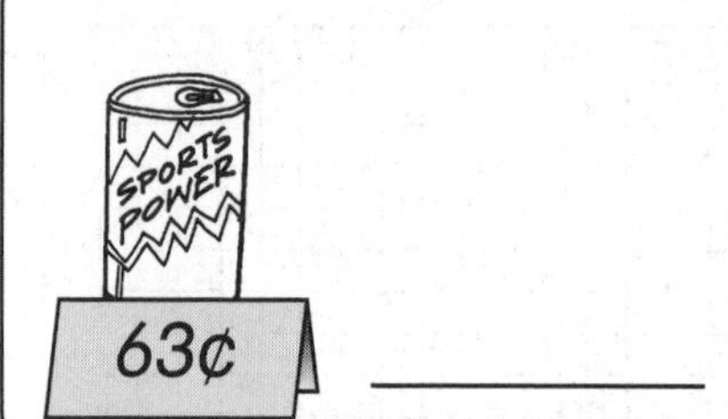

4. Write your own multiplication story problems about the drink containers.

Name Date

Multiplying Tens

Write the answers.

10×9	10×8	10×7	10×6	10×5	10×4
30×9	30×8	30×7	30×6	30×5	30×4
60×9	60×8	60×7	60×6	60×5	60×4

10×3	20×3	30×3	40×3	50×3	60×3

90×1	90×2	90×3	90×4	90×5	90×6

10×9	20×8	30×7	40×6	50×5	60×4

These examples provide practice in multiplying "tens" by 1-digit numbers. When he or she has completed the page, ask your child to tell how the answers change across the first row. (They decrease by 10 each time.) Repeat the question for other rows. Encourage your child to look for other patterns in the answers.

Name Date

Encyclopedia Sale

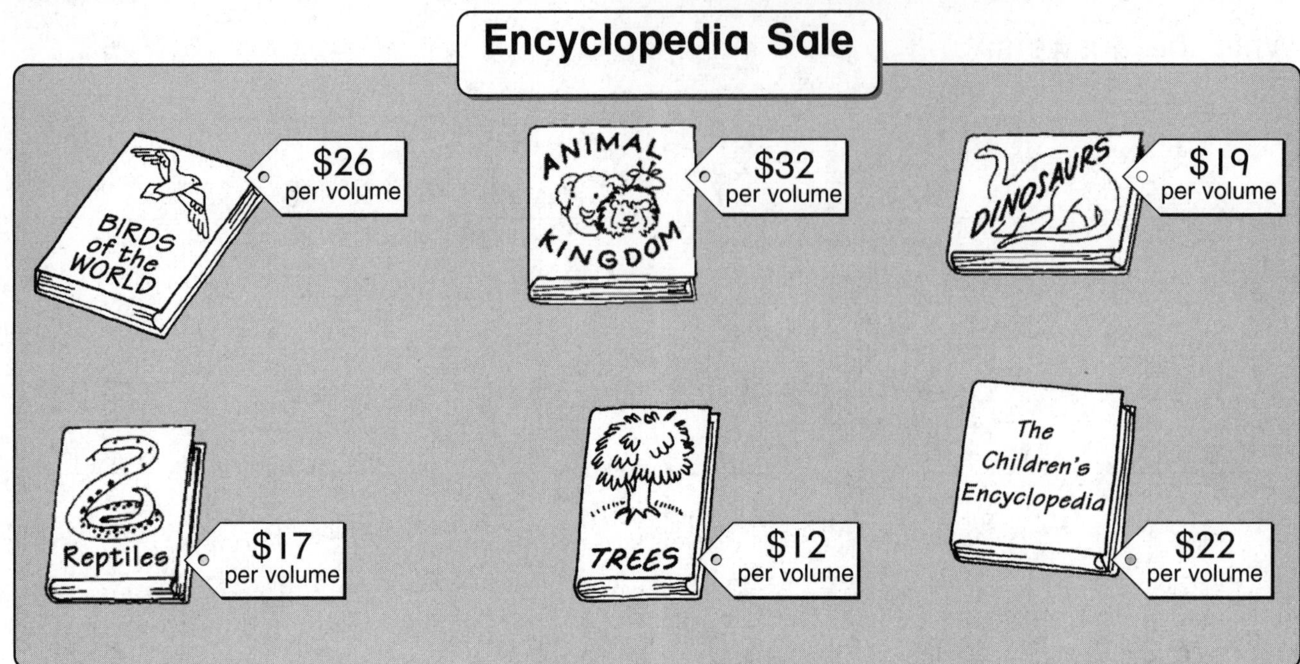

1. If you bought 3 volumes of ***Reptiles***, would they cost **more** or **less** than $50? ______________

2. Would 4 volumes of ***Trees*** cost more or less than $40? ______________

3. Ring the number of volumes of ***Dinosaurs*** you could buy with these certificates.

$40 gift certificate	1	2	3	4	5
$70 gift certificate	1	2	3	4	5
$110 gift certificate	1	2	3	4	5

4. How many volumes of ***Birds of the World*** could you buy with $80? _____

5. How many volumes of ***The Children's Encyclopedia*** could you buy with $100? _____

6. On the line below, place a ✓ next to the amount you think you would spend if you bought 9 volumes of ***Animal Kingdom***.

$260 $270 $280 $290 $300 $310 $320 $330 $340 $350

 Use with Investigation 7.5

Name Date

Finding Multiplication Patterns

1. Fill in the answers.

$3 \times 3 =$ ____ $2 \times 4 =$ ____	$4 \times 4 =$ ____ $3 \times 5 =$ ____	$5 \times 5 =$ ____ $4 \times 6 =$ ____
$6 \times 6 =$ ____ $5 \times 7 =$ ____	$7 \times 7 =$ ____ $6 \times 8 =$ ____	$8 \times 8 =$ ____ $7 \times 9 =$ ____

2. Write about the pattern in the answers in Question **1**.

____

____

____

3. Keep the pattern going. Use your calculator to help you find the answers.

$10 \times 10 =$	*100*	$= 100 -$	*0*	$20 \times 20 =$ ____	$= 400 -$ ____
$9 \times 11 =$	____	$= 100 -$	*1*	$19 \times 21 =$ ____	$= 400 -$ ____
$8 \times 12 =$	____	$= 100 -$	*4*	$18 \times 22 =$ ____	$= 400 -$ ____
$7 \times 13 =$	____	$= 100 -$	____	$17 \times 23 =$ ____	$= 400 -$ ____
$6 \times 14 =$	____	$= 100 -$	____	$16 \times 24 =$ ____	$= 400 -$ ____
$5 \times 15 =$	____	$= 100 -$	____	$15 \times 25 =$ ____	$= 400 -$ ____
$4 \times 16 =$	____	$= 100 -$	____	$14 \times 26 =$ ____	$= 400 -$ ____
$3 \times 17 =$	____	$= 100 -$	____	$13 \times 27 =$ ____	$= 400 -$ ____
$2 \times 18 =$	____	$= 100 -$	____	$12 \times 28 =$ ____	$= 400 -$ ____

4. Write about the pattern you created in Question **3**.

____

____

____

Name Date

Multiplying 2-digit Numbers

Write the answers. For each group, use the first 2 answers to help figure out the third answer.

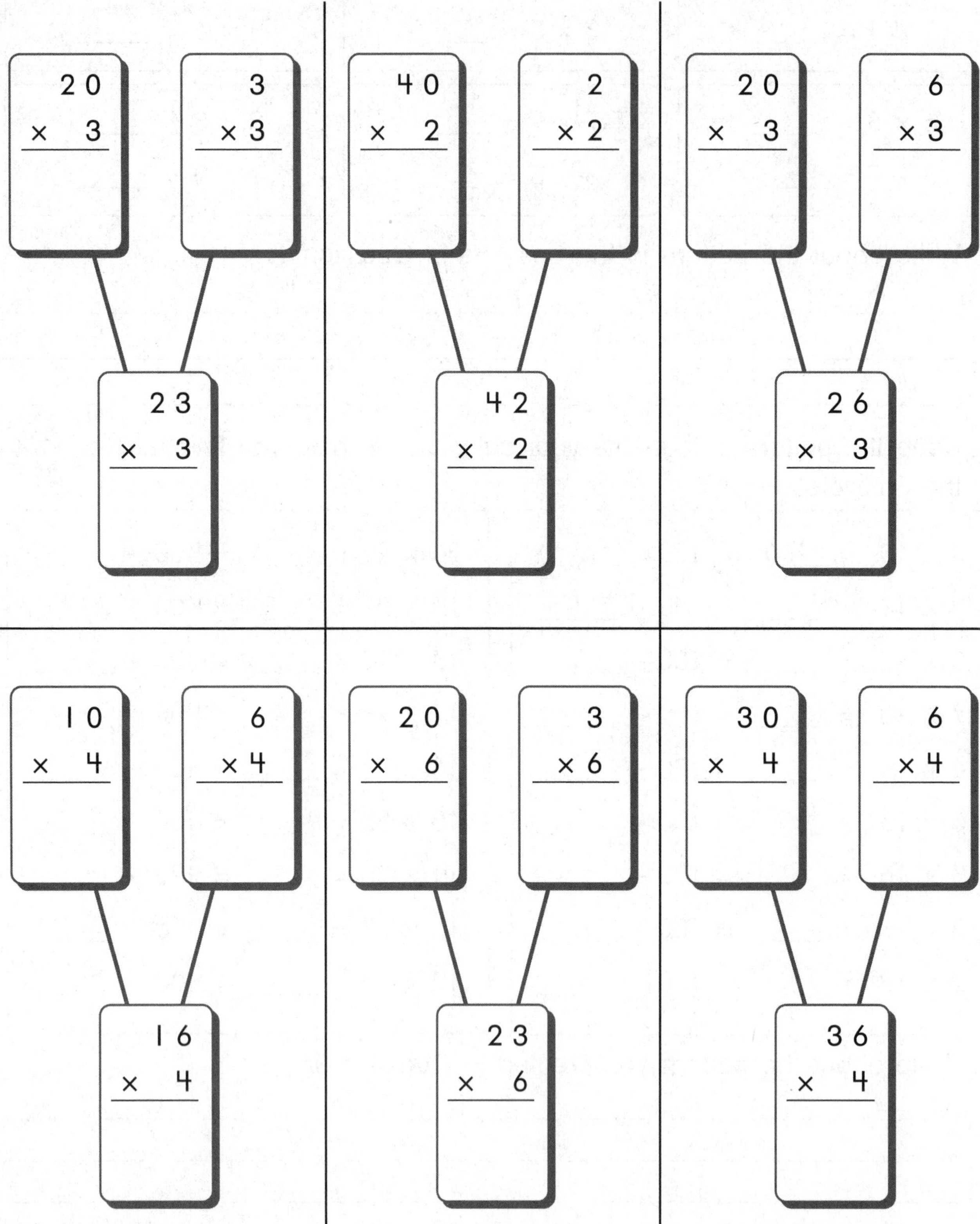

Name Date

Using a Calendar

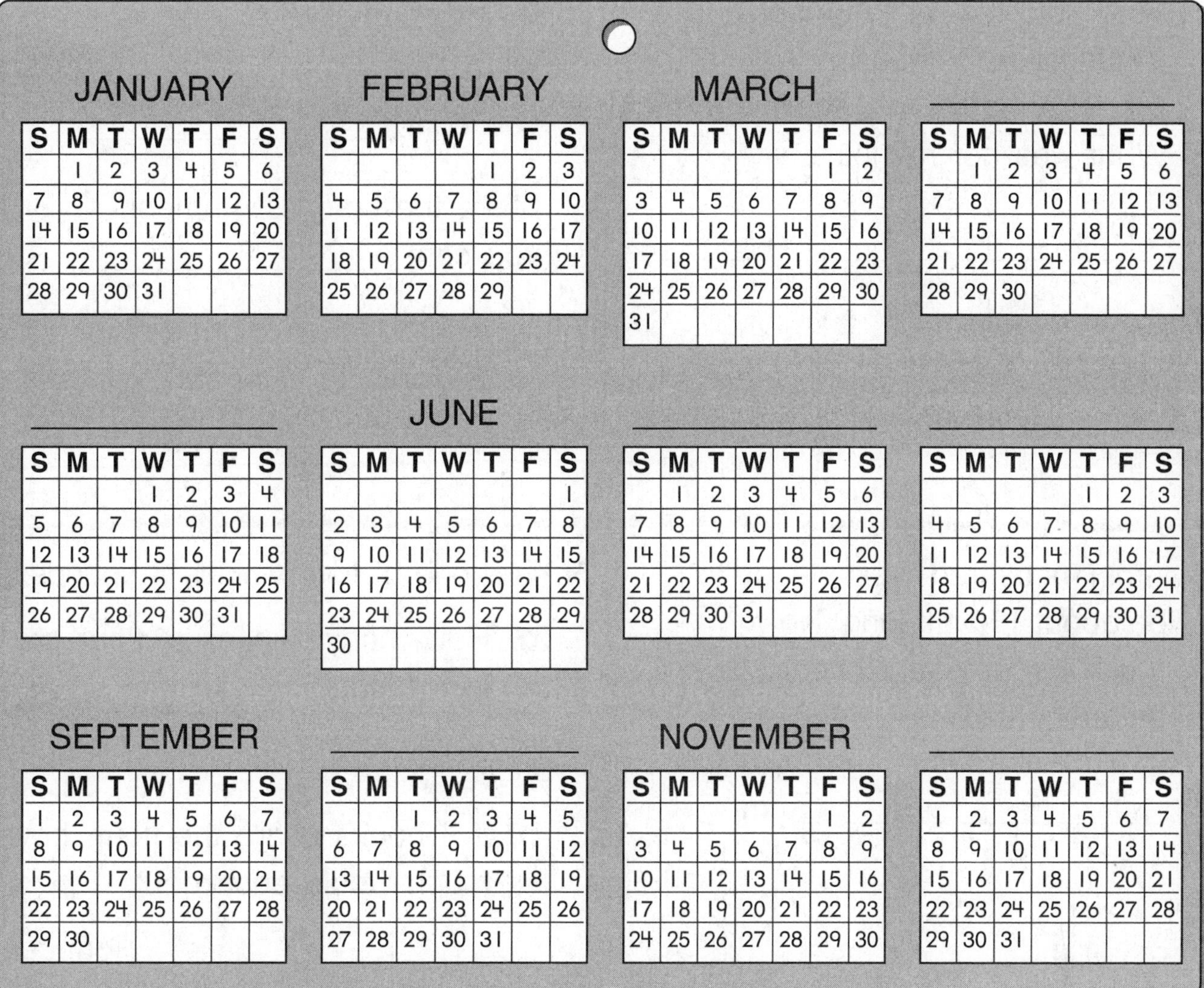

JANUARY

S	M	T	W	T	F	S
	1	2	3	4	5	6
7	8	9	10	11	12	13
14	15	16	17	18	19	20
21	22	23	24	25	26	27
28	29	30	31			

FEBRUARY

S	M	T	W	T	F	S
				1	2	3
4	5	6	7	8	9	10
11	12	13	14	15	16	17
18	19	20	21	22	23	24
25	26	27	28	29		

MARCH

S	M	T	W	T	F	S
					1	2
3	4	5	6	7	8	9
10	11	12	13	14	15	16
17	18	19	20	21	22	23
24	25	26	27	28	29	30
31						

____

S	M	T	W	T	F	S
	1	2	3	4	5	6
7	8	9	10	11	12	13
14	15	16	17	18	19	20
21	22	23	24	25	26	27
28	29	30				

____

S	M	T	W	T	F	S
			1	2	3	4
5	6	7	8	9	10	11
12	13	14	15	16	17	18
19	20	21	22	23	24	25
26	27	28	29	30	31	

JUNE

S	M	T	W	T	F	S
						1
2	3	4	5	6	7	8
9	10	11	12	13	14	15
16	17	18	19	20	21	22
23	24	25	26	27	28	29
30						

____

S	M	T	W	T	F	S
	1	2	3	4	5	6
7	8	9	10	11	12	13
14	15	16	17	18	19	20
21	22	23	24	25	26	27
28	29	30	31			

____

S	M	T	W	T	F	S
				1	2	3
4	5	6	7	8	9	10
11	12	13	14	15	16	17
18	19	20	21	22	23	24
25	26	27	28	29	30	31

SEPTEMBER

S	M	T	W	T	F	S
1	2	3	4	5	6	7
8	9	10	11	12	13	14
15	16	17	18	19	20	21
22	23	24	25	26	27	28
29	30					

____

S	M	T	W	T	F	S
		1	2	3	4	5
6	7	8	9	10	11	12
13	14	15	16	17	18	19
20	21	22	23	24	25	26
27	28	29	30	31		

NOVEMBER

S	M	T	W	T	F	S
					1	2
3	4	5	6	7	8	9
10	11	12	13	14	15	16
17	18	19	20	21	22	23
24	25	26	27	28	29	30

____

S	M	T	W	T	F	S
1	2	3	4	5	6	7
8	9	10	11	12	13	14
15	16	17	18	19	20	21
22	23	24	25	26	27	28
29	30	31				

1. Write the missing month names on the calendar.
2. How do you know this is the calendar of a leap year?

   ____

3. Look at the calendar to find which day of the week is the first day of

   Summer ______________ Fall ______________

   Winter ______________ Spring ______________

4. Use these colors to show the dates on your calendar.
   - Blue – New Year's Day (Jan. 1)
   - Red – Independence Day (July 4)
   - Orange – Leap Year Day (Feb. 29)
   - Pink – Memorial Day (May 30)
   - Green – Thanksgiving (Nov. 28)
   - Yellow – Your birthday
5. Color all the days when students would not be at school.
6. How many days on the calendar would be school days? ______________

Name Date

Maintaining Concepts and Skills

1. Vera said that her number had 2 hundreds, 5 thousands, 6 ones, and 9 ten thousands. What was her number?

   ____

2. Use 8, 0, 1, and 9.

   Make the largest number you can. ____________

   Make the smallest number you can. ____________

   Find the difference. ________

3. Amin measured the table top to be about 140 cm long. How much longer would it be if it was 2 meters long?

   ______________ cm

4. The bag of flour weighs 3 pounds.

   a. How many ounces of flour are in the bag?

   ______________ ounces

   b. If 4 cooks share the flour, how much would each get?

   ______________ ounces

5. A school's play area is a large rectangle. Draw 2 different ways that 3 classes can have equal shares.

6. Write three numbers that are greater than 10,000 and less than 12,000.

   ________ ________ ________

Name Date

Calculating Time

1. Write the number of minutes that have **passed** since the hour. Then write how many minutes there are **to** the next hour.

Minutes past the hour		Minutes to the hour
________________		________________
________________	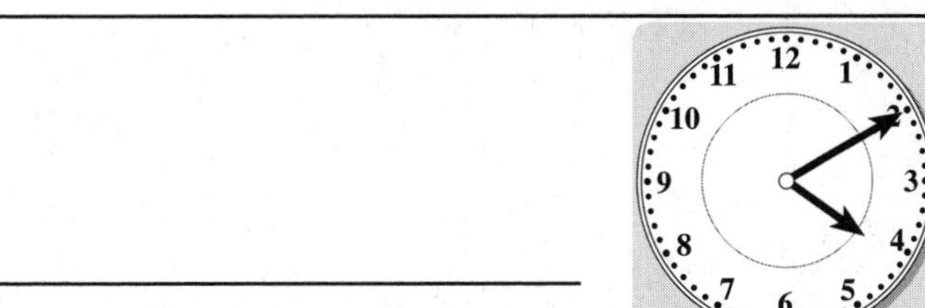	________________
________________	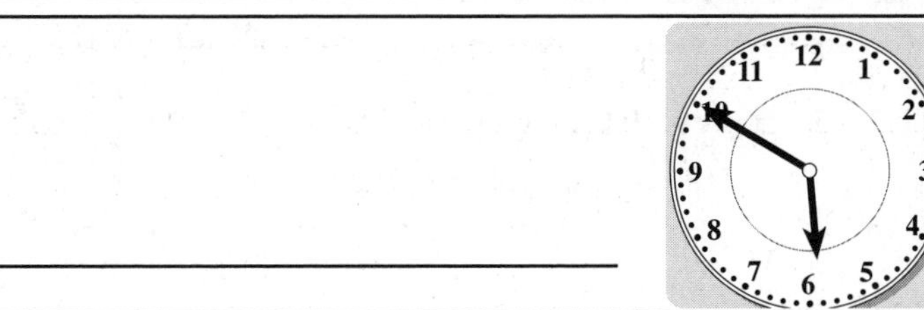	________________
________________	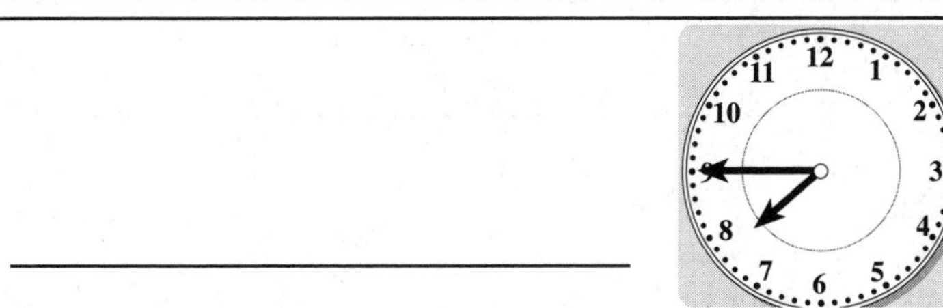	________________
________________		________________

2. Write these times.

___ minutes to ___

___ minutes to ___

___ minutes to ___

___ minutes to ___

Name Date

Maintaining Concepts and Skills

1. Todd keeps his animal cards in a book. He puts 5 cards on a page. How many pages will he use to hold 37 new cards?

   ______________ pages

2. Three students shared 10 sets of stickers. How many sets did each student get?

   __________________ sets

   There were 12 stickers in each set. How many stickers did each student get?

   __________________ stickers

3. Write 3 fractions that are less than one-half.

   ________ ________ ________

4. Write 3 fractions that are greater than one-half.

   ________ ________ ________

5. Three-fourths of the class went on a bus trip. How many students would that be if the class had

   12 students? ____________

   20 students? ____________

   32 students? ____________

6. About how many kilograms are there in 4,900 grams?

   __________ kilograms

Name Date

Creating a Schedule

Show each event below in the correct place on the schedule. Use a different color for each event and write its starting time. Remember to use a.m. and p.m.

1. Morning recess starts at 10:30 a.m. and finishes at 10:45 a.m. Lunch recess starts at 12:00 noon and finishes at 12:45 p.m. How many minutes break are there in all? __________ minutes
2. Math starts at 9:45 a.m. and finishes at morning recess. How many minutes does it last? __________ minutes
3. Science starts after morning recess and goes for 30 minutes. What time does Science finish? __________
4. Students read quietly for **10** minutes before lunch recess. At what time do they start reading? __________
5. How many minutes are free for Social Studies in the session between morning recess and lunch recess?

   __________ minutes
6. School finishes at 3:00 p.m. The afternoon session is divided into 2 parts. Show and write the times when students would do Language Arts and Arts and Crafts.

School Schedule

8:30 a.m.

9:00 a.m.

10:00 a.m.

Name Date

Using Multiplication

This chart shows the number of balls that ball machines can make.

Football	Baseball	Basketball	Tennis ball
1 every 10 minutes	3 every 10 minutes	5 every 10 minutes	8 every 10 minutes

Suppose the machines work nonstop.

1. How many footballs could be made in

1 hour? ____________________

12 hours? ____________________

1 day? ____________________

2. How many baseballs could be made in

1 hour? ____________________

12 hours? ____________________

1 day? ____________________

3. How did you figure out the answers to Question 2?

____

4. If one machine can make 8 tennis balls in 10 minutes, how many would be made every hour by

2 machines? ____________________

3 machines? ____________________

4 machines? ____________________

5. How long would it take the basketball machine to make

20 basketballs? ____________________

100 basketballs? ____________________

200 basketballs? ____________________

Name Date

Relating Units of Time

Multiply. Write the answers.

30 × 1	30 × 2	30 × 3	30 × 4	30 × 5	30 × 6
60 × 1	60 × 2	60 × 3	60 × 4	60 × 5	60 × 6

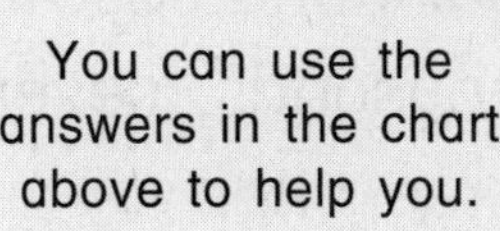

You can use the answers in the chart above to help you.

Write the missing numbers.

2 hours = ______ minutes	4 hours = ______ minutes
$1\frac{1}{2}$ hours = ______ minutes	$4\frac{1}{2}$ hours = ______ minutes
3 minutes = ______ seconds	$6\frac{1}{2}$ minutes = ______ seconds
$2\frac{1}{2}$ minutes = ______ seconds	$5\frac{1}{2}$ minutes = ______ seconds
300 minutes = ______ hours	30 minutes = ______ hour
360 seconds = ______ minutes	210 seconds = ______ minutes

Name Date

Maintaining Concepts and Skills

1. Draw shapes to show these fractions.

$\frac{1}{3}$ $\frac{2}{3}$ $\frac{1}{5}$ $\frac{4}{5}$

2. Which is more, $\frac{5}{8}$ of a pizza or $\frac{3}{4}$ of the same size pizza? Use drawings to show the answer.

____

3. Write the number from these mixed-up words.

four hundreds, five ten thousands, seven tens, three ones

____

4. What unit would you use to measure the amount of water you would put in a tub to take a bath?

____

5. Jon said the height of his fence was one and one-fourth meters. Sentha measured the same fence and said it was 128 centimeters high. Are they both right?

____

Explain your answer.

____

____

____

6. Three students shared 2 bars of modeling clay. At another table, 6 students shared 4 bars of modeling clay. Which group of students got more clay? Use words or pictures to show your answers.

____

____

Name Date

Maintaining Concepts and Skills

1. Jasmine said a number had 32 tens. Latima said it had 2 tens. Jasmine said the same number had 13 hundreds. Latima said it had 3 hundreds. Jasmine said the number had 9 ones. Latima agreed with her. Write a number that would make both of them correct.

____

2. Think of 2 numbers that could match these clues.

- The two numbers have the same digit in the tens place.
- The first number has the same digit in the hundreds place as the second number has in the ones place.
- All the digits in the first number are different.
- The numbers could be

________ and ________.

3. Four robins shared 18 worms equally. How many worms did each robin get?

____________ worms

4. Estimate the answers.

$63 \div 11 =$ ______

$130 \div 11 =$ ______

Explain your reasoning.

____

____

5. Roland ate $\frac{1}{3}$ of the pizza at his table. Talente ate $\frac{1}{2}$ of the pizza at his table. Roland said he ate more pizza than Talente. Make a drawing to show how Roland could be right.

6. Complete the following.

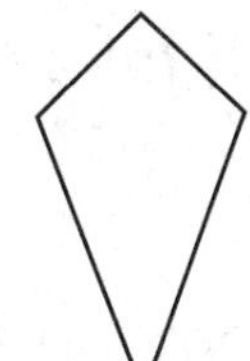

Number of sides ______

Number of angles ______

Perimeter ________ cm

Name ____________

Name Date

Using a Timetable

This timetable shows the times of flights from **Washington, D.C.**, to 4 other cities. (All are in the same time zone.)

Depart	Arrive	Flight No.	Stops	Depart	Arrive	Flight No.	Stops
To Atlanta				**To Miami**			
6:00 a.m.	8:15 a.m.	NS61	0	7:15 a.m.	10:30 a.m.	NS109	0
8:55 a.m.	11:10 a.m.	NS63	0	8:55 a.m.	1:45 p.m.	NS63	1
11:05 a.m.	1:20 p.m.	NS65	0	2:00 p.m.	6:45 p.m.	NS67	1
2:00 p.m.	4:15 p.m.	NS67	0	5:10 p.m.	9:50 p.m.	NS71	1
3:50 p.m.	6:05 p.m.	NS69	0	5:55 p.m.	9:20 p.m.	NS111	0
5:10 p.m.	7:25 p.m.	NS71	0				
7:55 p.m.	10:10 p.m.	NS73	0	**To New York**			
				6:00 a.m.	7:25 a.m.	NS202	0
To Boston				6:45 a.m.	8:10 a.m.	NS204	0
8:05 a.m.	9:35 a.m.	NS262	0	7:15 a.m.	8:40 a.m.	NS206	0
8:50 a.m.	10:20 a.m.	NS264	0	10:55 a.m.	12:20 p.m.	NS208	0
12:20 a.m.	1:45 p.m.	NS266	0	1:40 p.m.	3:05 p.m.	NS210	0
2:20 p.m.	3:45 p.m.	NS268	0	5:05 p.m.	6:30 p.m.	NS212	0
5:00 p.m.	6:25 p.m.	NS270	0	6:40 p.m.	8:05 p.m.	NS214	0
6:10 p.m.	7:35 p.m.	NS272	0	7:15 p.m.	8:40 p.m.	NS216	0

1. Write the departure times of the flights that leave

a. before 7:00 a.m. ______________________

b. between 5:30 p.m. and 6:30 p.m. ______________________

2. What time would you depart if you wanted to arrive in New York by 8:30 a.m? ______________

3. About how long does it take to travel from Washington, D.C. to Atlanta? ______________

How much **longer** does it take to travel nonstop from Washington, D.C. to Miami than to Atlanta? ______________

4. What time is the first afternoon flight to New York? ______________

5. What time would you need to leave Washington, D.C. if you want to meet a friend at the Boston airport at 6:30 p.m? ______________

6. Estimate the amount of time you would save by taking the 5:55 p.m. nonstop flight to Miami rather than the 5:10 p.m. flight. ______________

Name Date

Calculating Time Differences

HOME LINK

Write how long each bus trip took.

All the times are in the afternoon.

Starting time	Stopping time	Travel time
		The bus trip took ______ ______.
		The bus trip took ______ ______.
		The bus trip took ______ ______.
5:00	6:45	The bus trip took ______ ______.
4:15	5:45	The bus trip took ______ ______.
3:45	6:00	The bus trip took ______ ______.

Your child may wish to use a clock face to help him or her work with the digital times. The travel times may be written in any suitable form, such as 1 hour and 30 minutes, 90 minutes, or $1\frac{1}{2}$ hours.

Name Date

Prism Patterns

Prism Shape	Number of FACES	Number of VERTICES	Number of EDGES
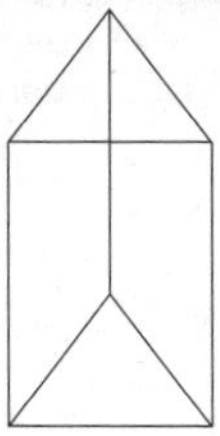 ______			
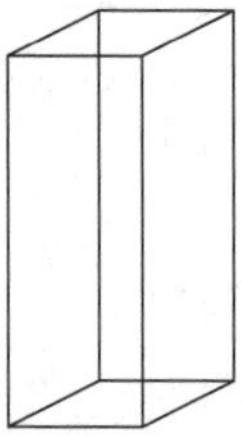 ______			
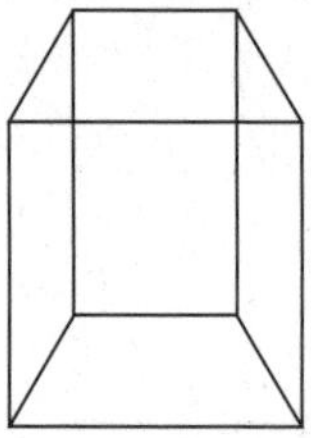 ______			
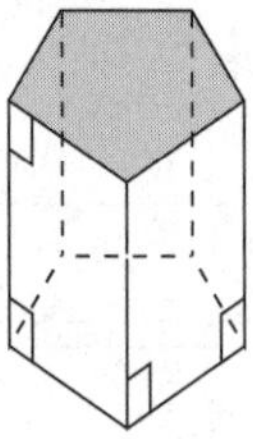 ______			
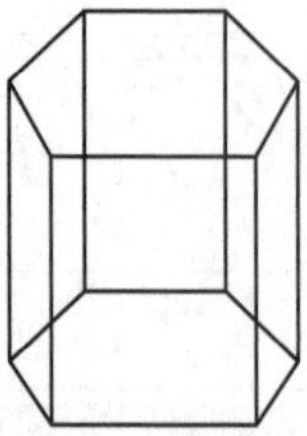 ______			

Name Date

Maintaining Concepts and Skills

1. Shade $\frac{3}{5}$ of the set.

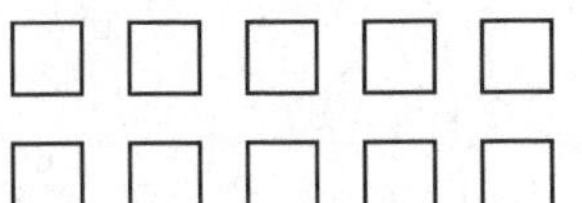

2. Shade $\frac{2}{3}$ of the set.

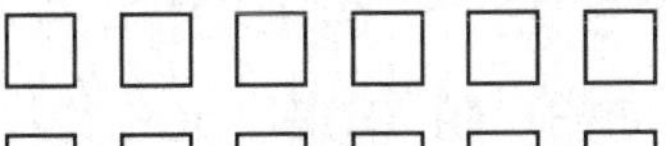

3. How many centimeters are in a meter?

_______________ cm

4. How many meters are in a kilometer?

______________________ m

5. Draw a 5-sided shape that has a perimeter of 20 centimeters.

6. Draw 3 line segments. The sum of their lengths should be 10 cm. Use fractions to tell what part of the total each line is.

Name Date

Finding Perimeter

Mrs. Gordan's fourth-grade class built a model of the Washington Monument. The measurements of one of the 4 faces are shown.
Find the perimeter of one face of the monument and the perimeter of the square base.

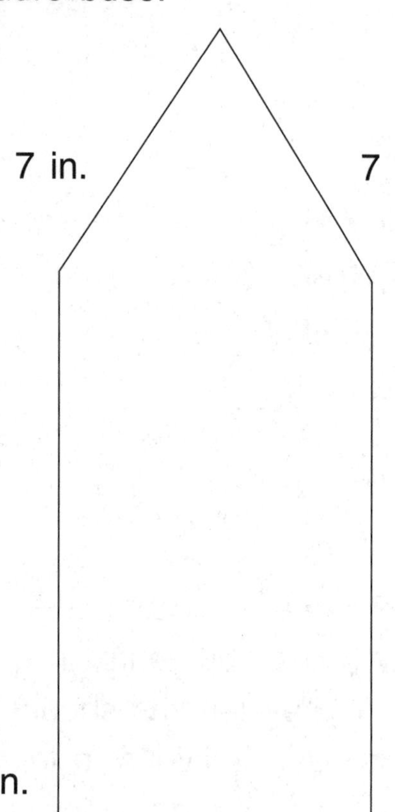

Perimeter of one face __________ in.

Perimeter of the base __________ in.

Name Date

Maintaining Concepts and Skills

1. Four students shared a box of pencils. What fraction of the box did each student get?

____

There were 12 pencils in the box. How many pencils did each student get?

________ pencils

2. Draw and shade a shape to show $\frac{3}{5}$.

3. Ring the unit you would use to measure the length of a car.

mile inch foot gallon

4. About how many pounds are there in 100 ounces? ________

How did you decide?

____

____

5. Write the numbers.

forty-three thousand, two hundred twenty-two ________

fifteen thousand, fifteen ________

fifteen hundred fifteen ________

What is another way to write the words for the last example?

____

____

6. Write the words for a number that has 13 thousands and 45 tens.

____

____

Name Date

Estimating and Finding Perimeter

1. Find objects listed in the chart below. Write your estimate of each perimeter. Follow these steps to measure each perimeter.

STEP 1: Use string to measure around the object.

STEP 2: Lay the string out flat and measure its length.

Object	My estimated perimeter	Measured perimeter
A. Math book		
B. Desktop		
C. Chalkboard eraser		
D. Chalkbox or shoebox		

2. Which object in the chart has the **shortest** perimeter?

____

Find an object in your classroom that you think has a **shorter** perimeter. Write your estimate. ______________

Measure it. Write the measured perimeter. ______________

3. Which object in the chart has the **longest** perimeter?

____

Find an object in your classroom that you think has a **longer** perimeter. Write your estimate. ______________

Measure it. Write the measured perimeter. ______________

4. List all the objects you measured **in order**, from the shortest perimeter to the longest perimeter.

____

____

Name Date

Maintaining Concepts and Skills

1. Write these 3 numbers.

a. the greatest number between 10,000 and 11,000 __________

b. the least number between 10,000 and 11,000 __________

c. the number halfway between the 2 numbers __________

2. Make a drawing to show a fraction that is close to one whole. Write the fraction next to your drawing.

3. About how many liters are in 5,280 milliliters?

__________________ liters

4. Write in words. 5,106

____

____

5. Write a 4-digit number. __________
Write a number riddle for your number. Use clues such as "the tens digit is greater than the hundreds digit."

____

____

____

____

6. Make a drawing to show $\frac{1}{3}$ and $\frac{2}{6}$.

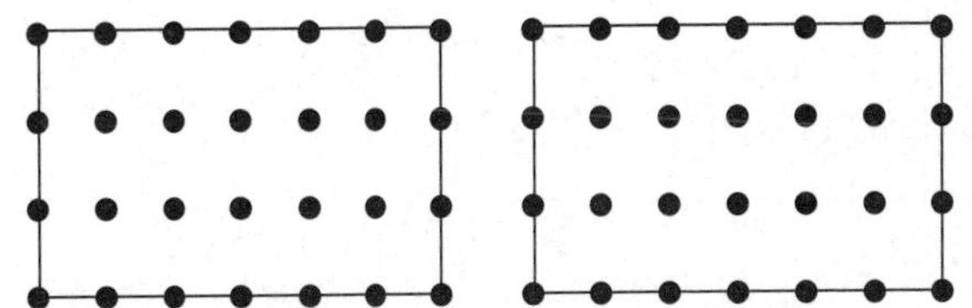

Does your drawing show that $\frac{1}{3}$ and $\frac{2}{6}$ are the same amount?

Explain your answer.

____

____

Name Date

Finding Perimeter

Make these shapes on a Geoboard.

- All the rectangles that have a perimeter of **16** units
- Other shapes that have a perimeter of **16** units

Copy each shape onto this dot paper.

Name Date

Calculating Perimeter

1. Measure one side of this square.
Fill in the first 2 spaces on the chart.
Now calculate the perimeter of the square.

Regular polygon	Length of one side	Number of sides	Perimeter
square	_____ cm	_______	_______

How did you calculate the perimeter? ____________________

____

2. Fill in the first 2 spaces on this chart. Then calculate the perimeter of the regular polygons shown below.

Regular polygon	Length of one side	Number of sides	Perimeter
hexagon	cm		
triangle	cm		
pentagon	cm		
octagon	cm		

Regular Polygon
A regular polygon is a shape with all sides and angles the same.

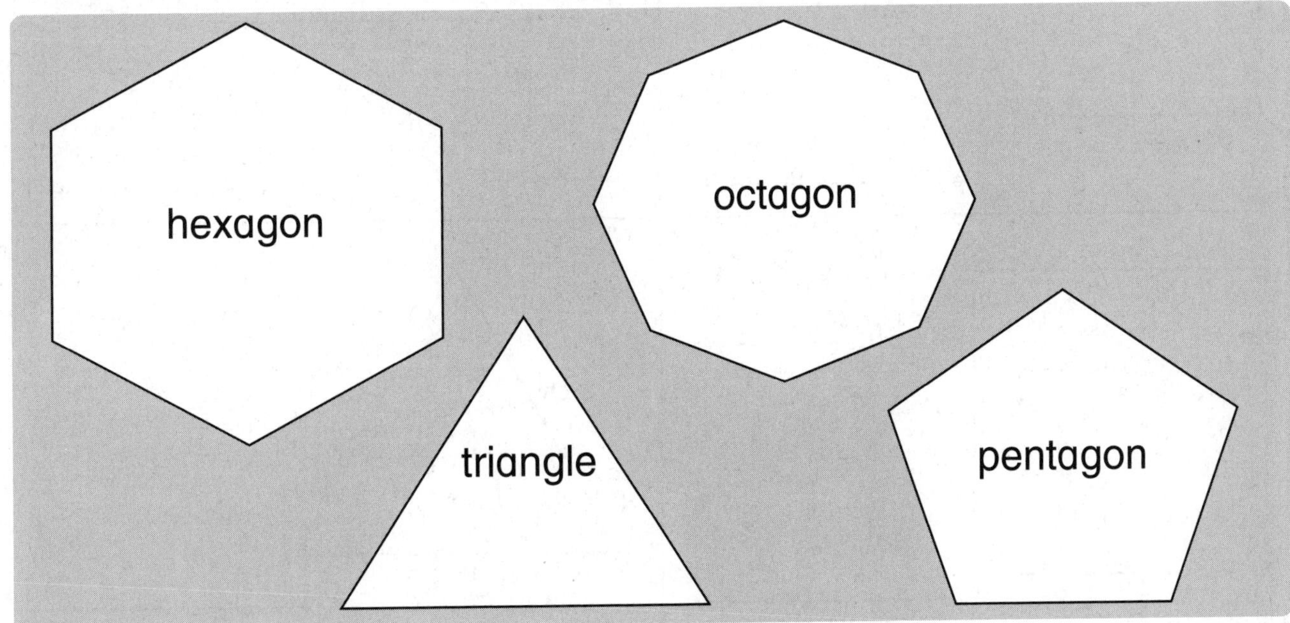

Name Date

Calculating the Perimeter of Polygons

Calculate the perimeter of each shape. Write the answer.

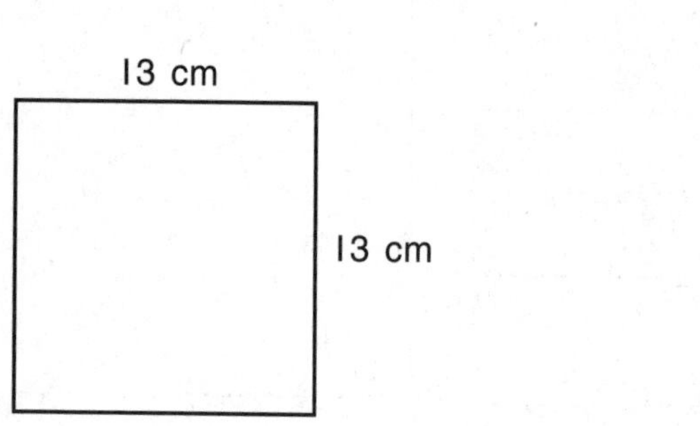

____

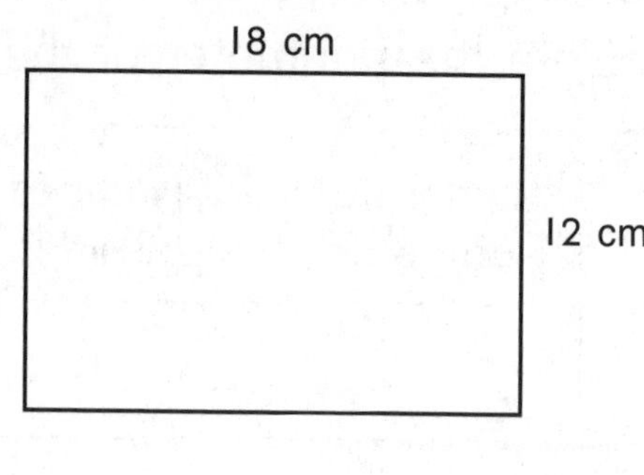

____

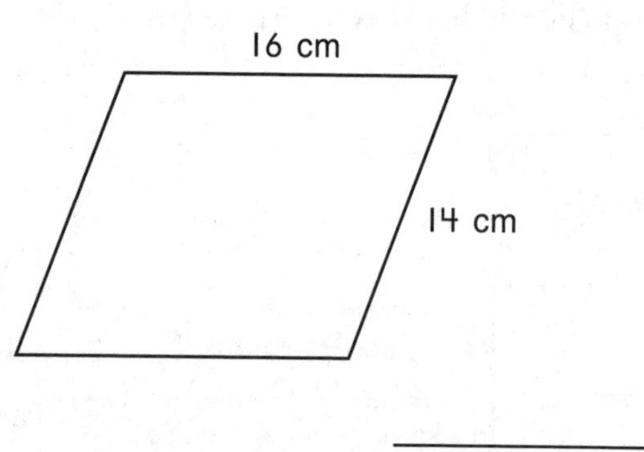

____

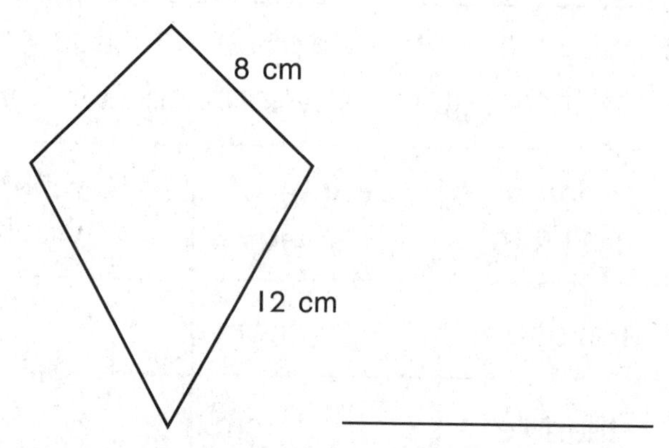

____

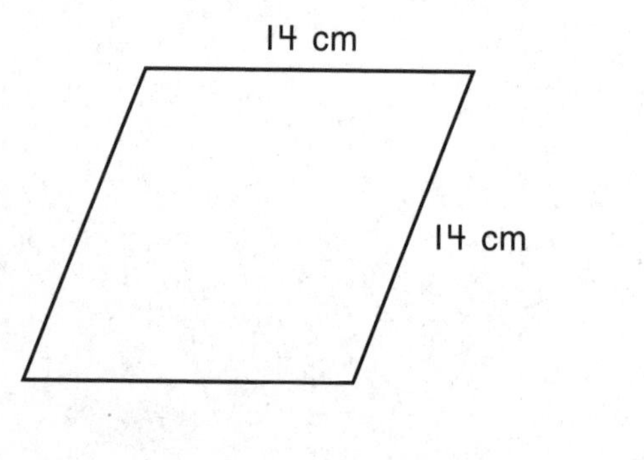

____

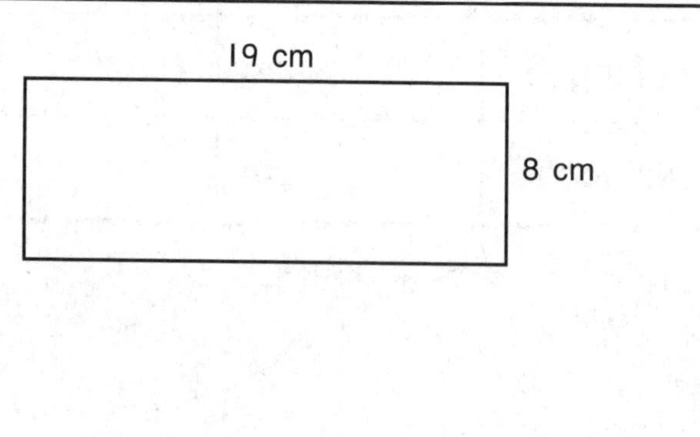

____

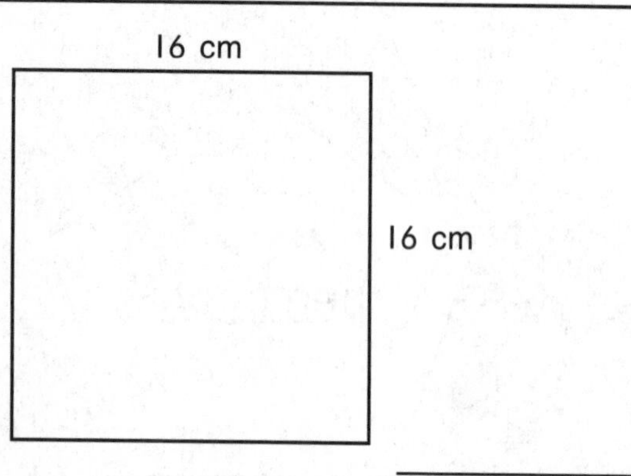

____

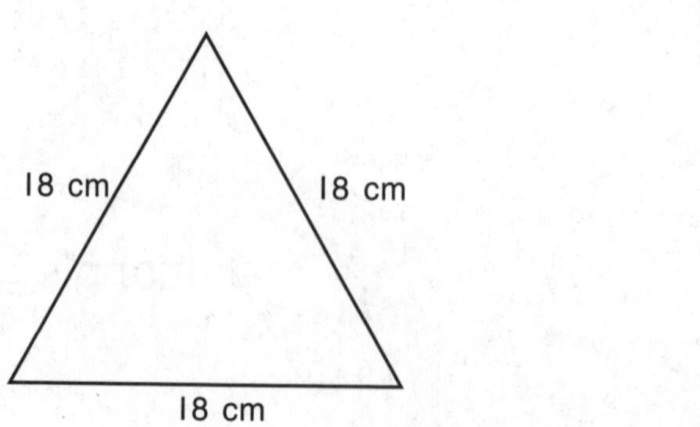

____

Use with Investigation 9.4

Name Date

Finding and Comparing Perimeter

Here are 3 ways of using ribbon to wrap the same package.

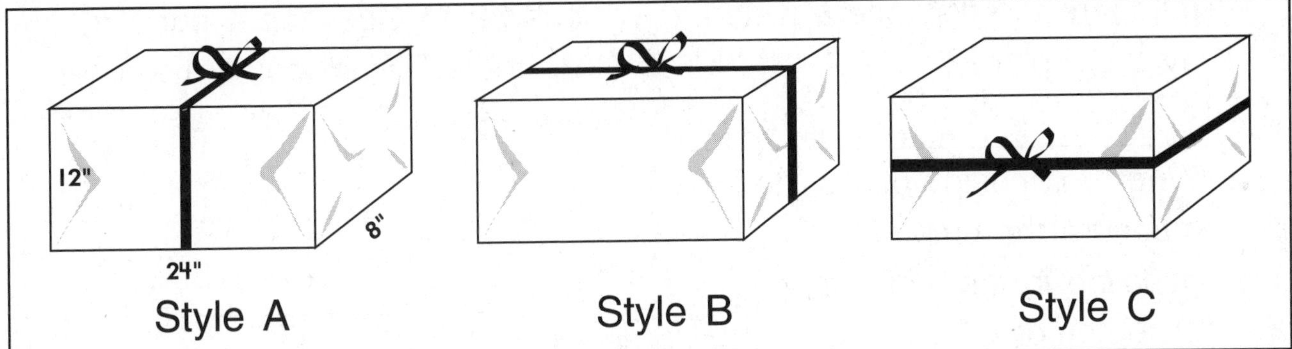

1. How much ribbon is needed for each style?
   (You will need an extra 8 inches for each knot.)

   A. ______________ B. ______________ C. ______________

2. How much ribbon is needed to wrap each of these in the 3 styles shown above? (Remember the 8 extra inches for the knot.)

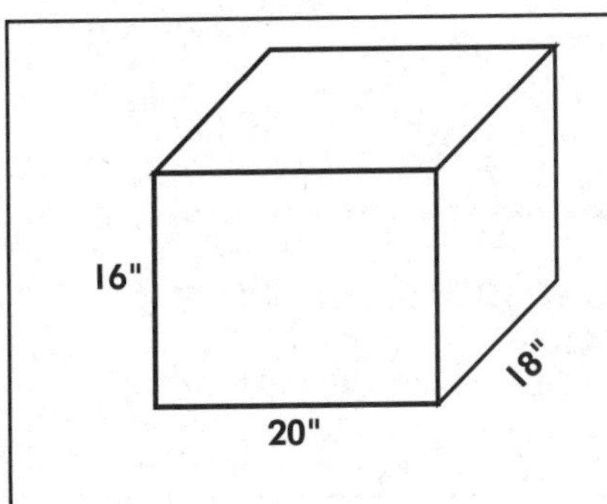

A. ______________

B. ______________

C. ______________

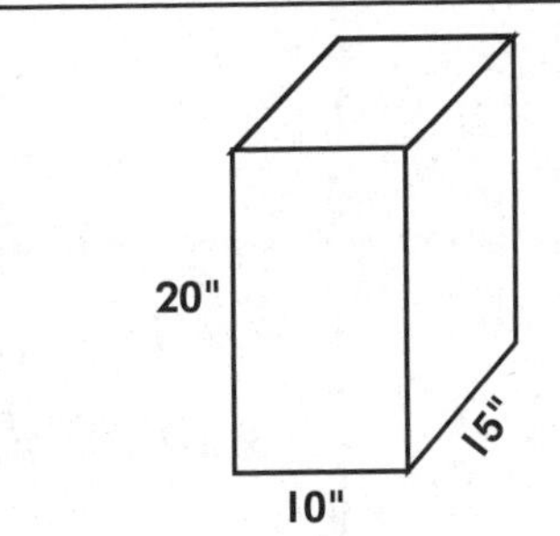

A. ______________

B. ______________

C. ______________

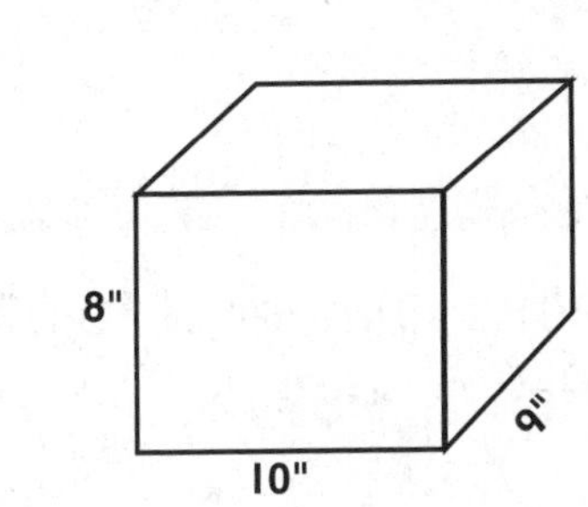

A. ______________

B. ______________

C. ______________

3. How much ribbon is needed to wrap each of these parcels?
   Add 8 inches for each knot.

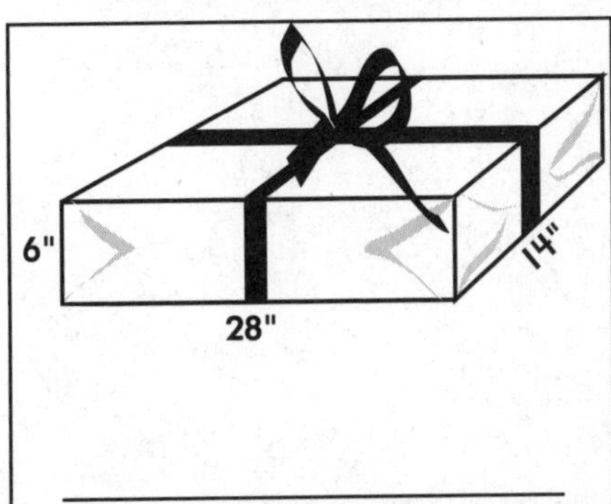

____

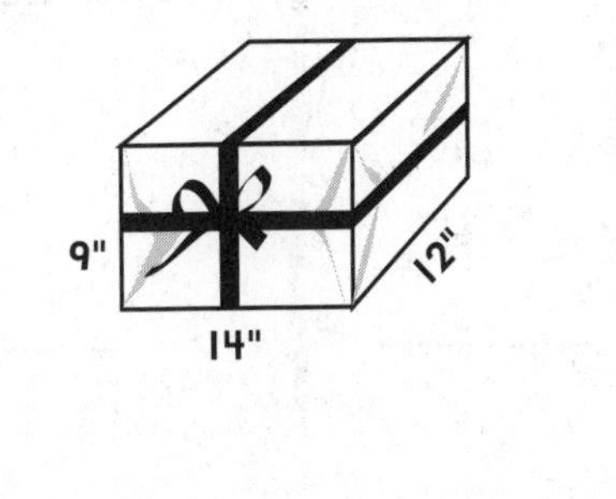

____

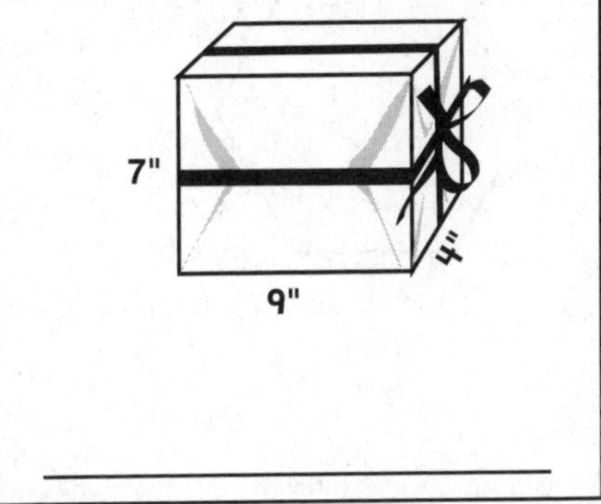

____

Name Date

Maintaining Concepts and Skills

1. a. What unit could you use to measure the distance traveled by an airplane?

____

b. What unit could you use to measure the amount of gasoline in the fuel tank of an automobile?

____

c. What unit could you use to measure the weight of a large dog?

____

2. Could there be more than one answer to Question **1** parts **a, b,** and **c**? What could the other answers be?

a. _______________

b. _______________

c. _______________

3. What number is 10,000 more than 2,995?

____

4. What number is 1,000 more than 98,345?

____

5. Use a centimeter ruler. Measure the perimeter of this shape.

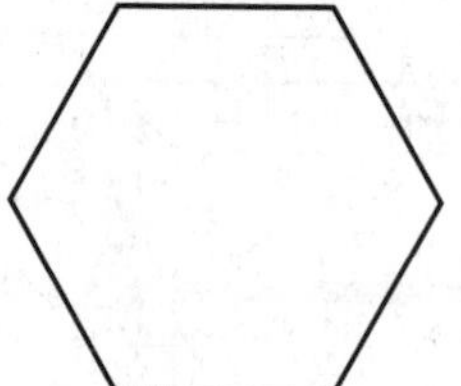

____

6. Begin at 0. Count by $\frac{1}{3}$ to 3. How many one-thirds did you use?

____

Name Date

Using Fractions

1. Shade the squares to show how the waffles were shared. Then write the fraction.

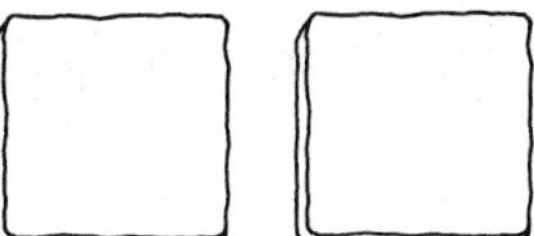

3 girls shared 2 waffles.
How much did each person get? ____________

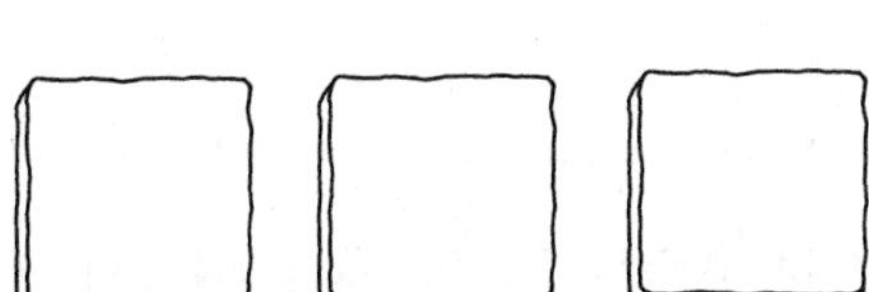

4 children shared 3 waffles.
How much did each person get? ________

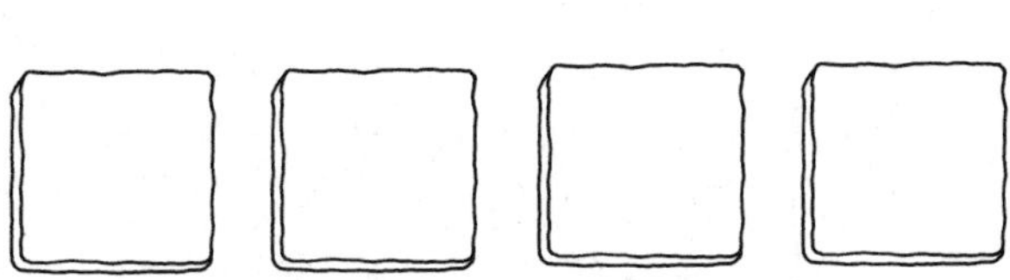

3 adults shared 4 waffles.
How much did each person get? ____________

2. Draw waffles to match these fraction problems.

4 children each got $1\frac{1}{2}$ waffles.

How many waffles were there in all? ____________

3 boys each got $\frac{2}{3}$ of a waffle.

How many waffles were there in all? ____________

5 family members each got $1\frac{2}{5}$ waffles.

How many waffles were there in all? ____________

Name Date

Maintaining Concepts and Skills

1. A flight departed from Atlanta at 6:05 a.m. It landed in Philadelphia 2 hours and 18 minutes later.
At what time did it land?

____

2. The concert starts at 8:00 p.m. If the trip to the concert takes 1 hour and 25 minutes, what time do you need to leave home?

____

3. Solve these.

$2 \times 12 =$ _____ $4 \times 12 =$ _____

$8 \times 12 =$ _____

Use the pattern to solve 16×12.

____

4. Use a pattern to help you solve.

$10 \times 17 =$ _____ $20 \times 17 =$ _____

$40 \times 17 =$ _____ $80 \times 17 =$ _____

5. Use the clues to draw the shape.
- I have four sides.
- Two sides are the same length.
- I have no right angles.
- Two of my sides are parallel.

6. a. Giyon can write her name in 18 seconds. How long will it take her to write her name 100 times?

____

b. How many times can Giyon write her name in 5 minutes?

____

7. Use the number tiles once each to make all the possible 3-digit numbers. Draw a box around the greatest number. Draw a ring around the least number.

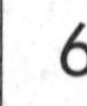

8

Name Date

Adding and Subtracting Fractions

Write the answers. You may draw a picture to help you.

$\frac{1}{2} + \frac{3}{4} = 1\frac{1}{4}$	$\frac{1}{4} + \frac{1}{4} =$ ______	$\frac{3}{4} + \frac{3}{4} =$ ______
$\frac{3}{4} + \frac{1}{4} =$ ______	$1\frac{1}{4} + \frac{1}{4} =$ ______	$1\frac{1}{2} + \frac{3}{4} =$ ______
$\frac{3}{4} - \frac{1}{4} =$ ______	$1\frac{3}{4} - \frac{1}{4} =$ ______	$1\frac{3}{4} - \frac{1}{2} =$ ______
$\frac{3}{4} - \frac{1}{2} =$ ______	$1\frac{1}{4} - \frac{1}{4} =$ ______	$1\frac{1}{4} - \frac{3}{4} =$ ______

Name Date

Adding and Subtracting Fractions

Write the answers. You may draw a picture to help you.

$\frac{1}{8} + \frac{3}{8} =$ ______	$\frac{1}{8} + \frac{5}{8} =$ ______	$\frac{1}{4} + \frac{3}{8} =$ ______
$\frac{1}{4} + \frac{5}{8} =$ ______	$\frac{3}{8} + \frac{7}{8} =$ ______	$1\frac{1}{8} + \frac{1}{8} =$ ______
$\frac{7}{8} - \frac{1}{8} =$ ______	$1 - \frac{3}{8} =$ ______	$\frac{7}{8} - \frac{2}{8} =$ ______
$\frac{7}{8} - \frac{5}{8} =$ ______	$1\frac{7}{8} - \frac{5}{8} =$ ______	$1\frac{3}{8} - \frac{5}{8} =$ ______

The students are using concrete models or pictures to help them add and subtract fractions. The fractions on this page can be shown by drawing a circle, square, or rectangle separated into 8 equal parts. If your child needs to work with a more concrete model, have him or her cut out 2 circles and fold and cut each into 8 equal parts.

Name Date

Solving Fraction Problems

Solve these fraction problems.

1. After a party, there were $2\frac{1}{3}$ cornbread squares left over.
   Shade enough cornbread squares to show about $2\frac{1}{3}$.

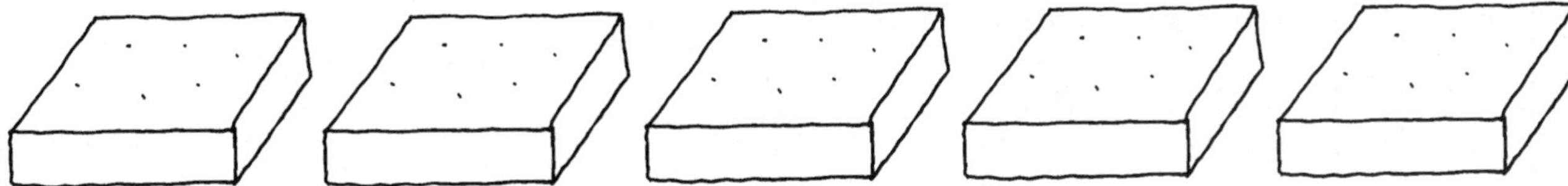

2. Arby ate $\frac{1}{2}$ of the Hawaiian pizza.
   Doc ate $\frac{1}{3}$ of the cheese pizza.
   Doc ate more pizza than Arby.
   Show how Doc could have eaten more than Arby.

Hawaiian pizza	Cheese pizza

Chocolate pie

3. When Sam got home,
   $\frac{1}{2}$ of a chocolate pie was left.
   Sam decided to eat $\frac{1}{2}$ of it.
   Draw a picture to show how much pie Sam ate.
   What fraction of the whole pie did Sam eat?

   ____

Name Date

Maintaining Concepts and Skills

1. Use this flight schedule to answer the questions. All flights are in the same time zone.

Flight No.	Departs	Arrives
168	8:30 a.m.	11:33 a.m.
344	1:05 p.m.	3:10 p.m.

a. How much longer is Flight 168 than Flight 344?

____

b. Flight 344 was delayed for 34 minutes. What time will it arrive?

____

2. Draw a rectangle. Make the sides 2 cm, 5 cm, 2 cm, and 5 cm.

3. Draw and shade a shape to show $\frac{3}{5}$.

4. Write the numbers for these.

a. twelve thousand, twelve ______

b. twelve hundred twelve ______

What is another way to write the words for **b**? _______________

____

5. Which is greater, $\frac{1}{2}$ or $\frac{1}{3}$? _______________

How do you know? _______________

____

Use 3 different ways to convince a friend which fraction is greater. You may use a drawing, a number line, change to a decimal, change to like numerators, or any other way that would help your friend to understand.

Name Date

Fractions on a Number Line

Mark an ✗ to show $\frac{1}{2}$.

Mark an ✗ to show $\frac{1}{3}$.

Mark an ✗ to show $\frac{3}{8}$.

Mark an ✗ to show $\frac{3}{4}$.

Mark an ✗ to show $\frac{5}{6}$.

Mark an ✗ to show $\frac{2}{5}$.

0 1

Name Date

Solving Fraction Problems

Solve these fraction problems. You can draw pictures to help.

1. How many halves are needed to make 2 wholes? _________

2. How many fourths are needed to make 4 wholes? _________

   How do you know? _______________

3. How many fourths are needed to make $2\frac{1}{2}$? _________

   How do you know? _______________

4. If you have 5 thirds, how many wholes can you make? ___________

   How much is left over? ___________

5. Greg has $3\frac{2}{3}$ yards of fancy trim. Cindy has $4\frac{2}{3}$ yards of fancy trim. How much fancy trim do they have in all? _________ yards

6. How many pints of paint does Tom need to paint his tree house? He needs $\frac{2}{3}$ pint to paint the inside and $1\frac{2}{3}$ pints to paint the outside.

   ____

   How do you know? _______________

Name Date

Using Number Lines Greater than 1

Mark an ✗ to show 1.

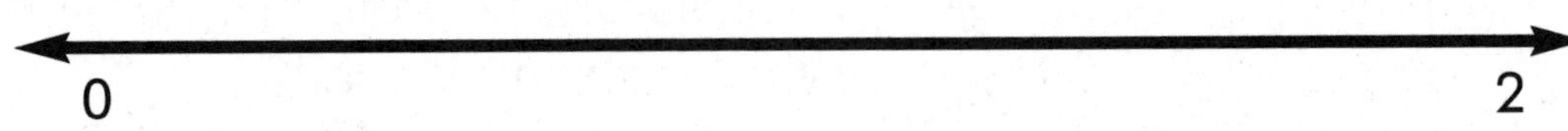

Mark an ✗ to show $\frac{1}{2}$.

Mark an ✗ to show $1\frac{1}{3}$.

Mark an ✗ to show $\frac{7}{8}$.

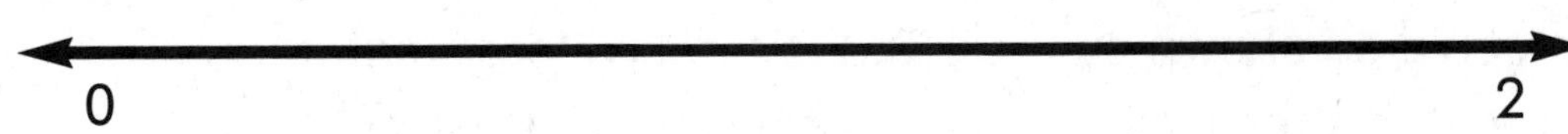

Mark an ✗ to show $1\frac{3}{4}$.

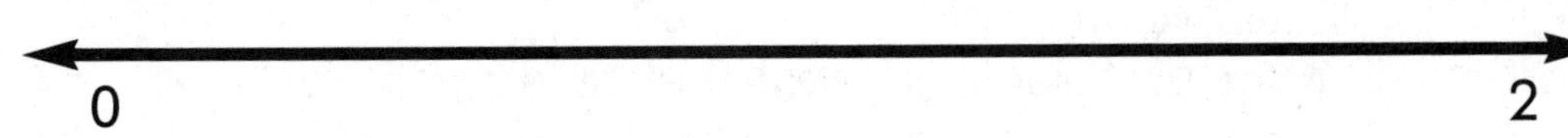

Mark an ✗ to show $1\frac{2}{3}$.

0 2

Name Date

Maintaining Concepts and Skills

1. Jamie ate $\frac{1}{3}$ of the pie. How much of the pie was left?

______

2. Allison cut chart paper into 5 parts. She gave one part to Danielle and 2 parts to Choi. What fraction was left?

______

3. How many kilograms are in 4,000 grams?

______ kilograms

4. Write the number.
ten thousand, four hundred six

______

5. Write the words for 1,093.

______

______

6. Write the words.
14,263 ______

______

12,001 ______

______

9,505 ______

______

7. What fraction could be shown by the shaded part of this rectangle?

______

8. How many centimeters is one-half of 3 meters?

______ cm

9. Shade the correct number of objects in each set to show $\frac{2}{5}$.

a. 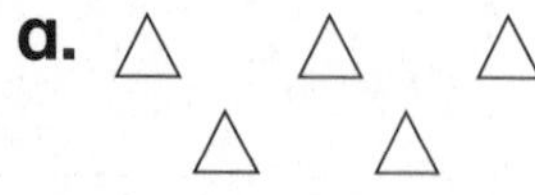

b. 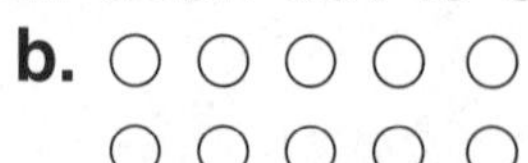

c. 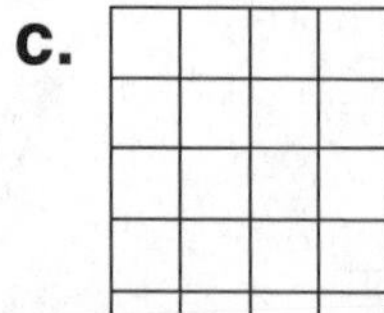

d. 

Name Date

Adding and Subtracting Fractions

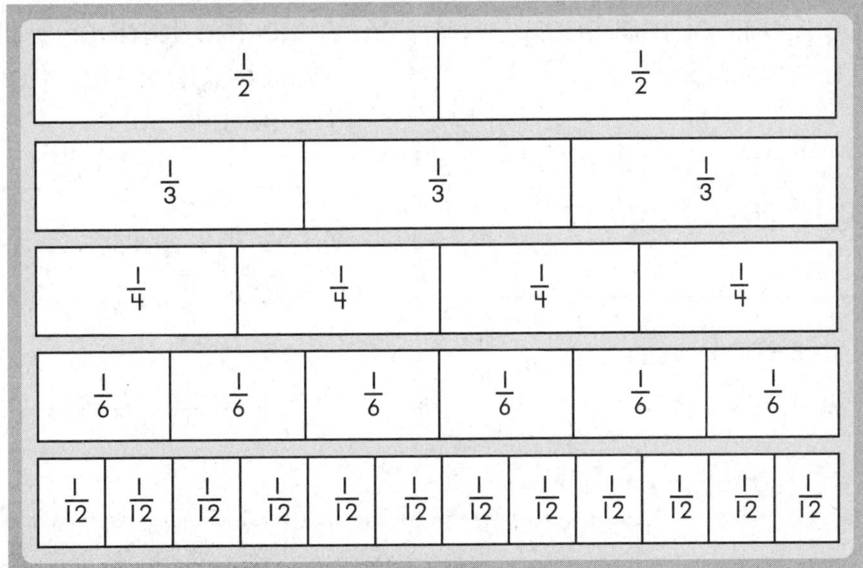

Use the fraction strips to help you add or subtract the fractions. Write the answers.

$\frac{1}{3} + \frac{1}{12} =$ ________	$\frac{1}{2} + \frac{1}{12} =$ ________
$\frac{1}{6} + \frac{1}{12} =$ ________	$\frac{1}{3} + \frac{1}{4} =$ ________
$\frac{7}{12} - \frac{1}{2} =$ ________	$\frac{11}{12} - \frac{1}{4} =$ ________
$\frac{2}{3} - \frac{1}{4} =$ ________	$\frac{5}{6} - \frac{1}{2} =$ ________

Name Date

Maintaining Concepts and Skills

1. Give the best unit of measure.
Length of a bookcase

____

Water in a large storage tank

____

Weight of a pencil

____

2. Write the largest 5-digit number you can that has no 2 digits the same.

____

Write the smallest 5-digit number you can that has no 2 digits the same.

____

3. What number is

100 less than 9,056? _______________

1,000 less than 56,001? _______________

4. Alta had $\frac{1}{2}$ of a quart of milk. Her recipe takes $\frac{1}{2}$ of a pint of milk. How many batches of the recipe can she make?

_______ batches

5. Draw a pentagon that has a perimeter of 10 cm. (Don't make all sides the same length.)

6. a. Anita's birthday is on the second Thursday of November.
What day is that? _______________

NOVEMBER

S	M	T	W	T	F	S
		1	2	3	4	5
6	7	8	9	10	11	12
13	14	15	16	17	18	19
20	21	22	23	24	25	26
27	28	29	30			

b. Jamal visited his grandmother 5 days before Anita's birthday. What day did he visit his grandmother?

____

Name Date

Adding and Subtracting Fractions

$\frac{1}{2}$	$\frac{1}{2}$

$\frac{1}{3}$	$\frac{1}{3}$	$\frac{1}{3}$

$\frac{1}{4}$	$\frac{1}{4}$	$\frac{1}{4}$	$\frac{1}{4}$

$\frac{1}{6}$	$\frac{1}{6}$	$\frac{1}{6}$	$\frac{1}{6}$	$\frac{1}{6}$	$\frac{1}{6}$

$\frac{1}{12}$	$\frac{1}{12}$	$\frac{1}{12}$	$\frac{1}{12}$	$\frac{1}{12}$	$\frac{1}{12}$	$\frac{1}{12}$	$\frac{1}{12}$	$\frac{1}{12}$	$\frac{1}{12}$	$\frac{1}{12}$	$\frac{1}{12}$

Use the fraction strips to help you add or subtract the fractions. Write the answers.

$\frac{1}{6} + \frac{5}{12} =$ ________

$\frac{2}{3} + \frac{1}{12} =$ ________

$\frac{5}{12} + \frac{1}{2} =$ ________

$\frac{2}{3} + \frac{1}{4} =$ ________

$\frac{11}{12} - \frac{1}{4} =$ ________

$\frac{11}{12} - \frac{1}{3} =$ ________

$\frac{3}{4} - \frac{1}{3} =$ ________

$\frac{5}{6} - \frac{1}{4} =$ ________

In class, students are using concrete or pictorial models to help them add and subtract fractions that have different denominators. The fraction strips will help your child see that $\frac{1}{3} + \frac{1}{12}$ is the same as $\frac{4}{12} + \frac{1}{12}$, and so on. Answers do not have to be written in simplest form at this stage. For example: $\frac{9}{12}$ and $\frac{3}{4}$ are equally acceptable.

Name Date

Reading and Writing Hundredths

Write the fraction and the decimal for the shaded part of each 100 grid.

1. 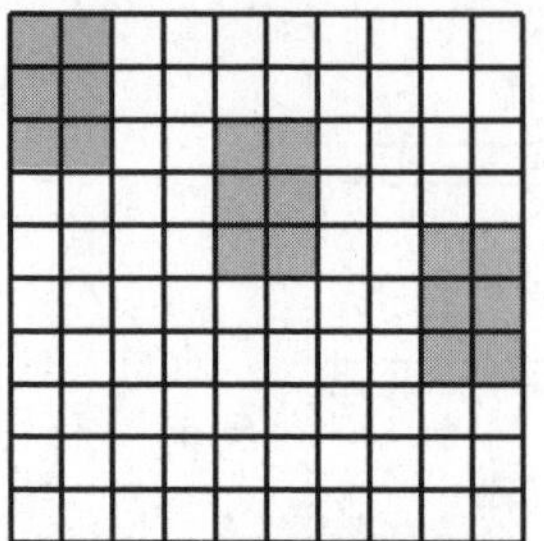

Fraction ______________

Decimal ______________

2. 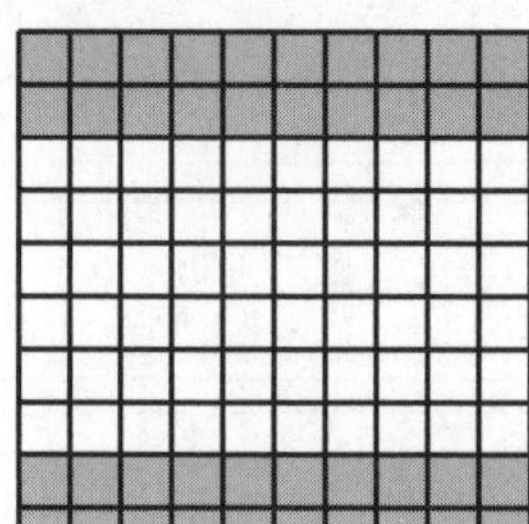

Fraction ______________

Decimal ______________

3. 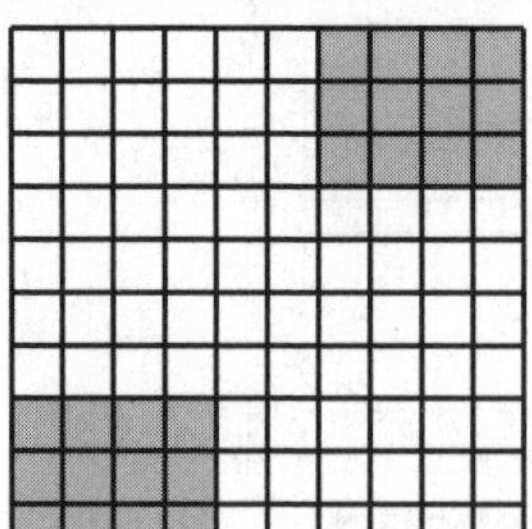

Fraction ______________

Decimal ______________

4. 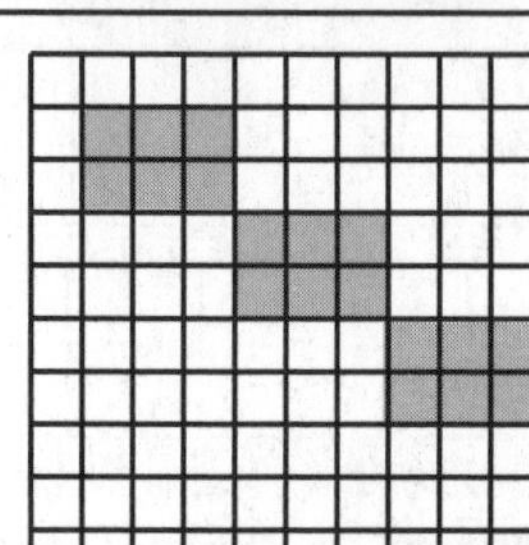

Fraction ______________

Decimal ______________

5. 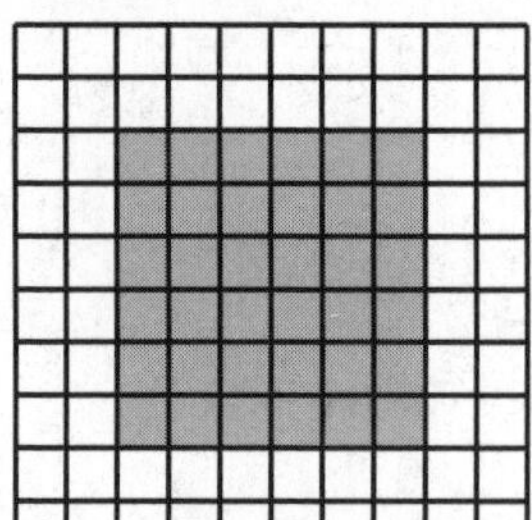

Fraction ______________

Decimal ______________

6. 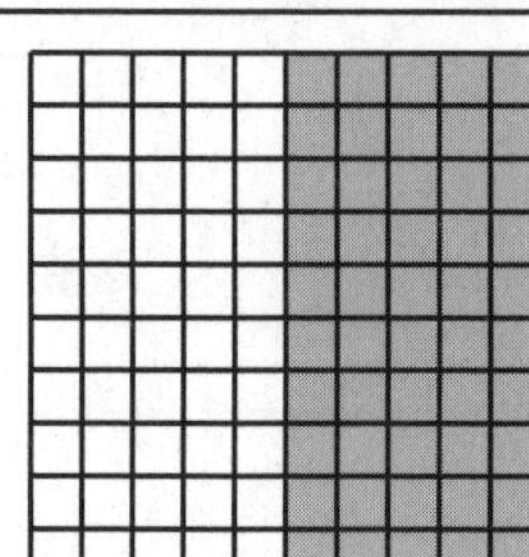

Fraction ______________

Decimal ______________

Shade in the 100 grid to represent the correct fraction or decimal. Then fill in the missing number for each problem.

1. 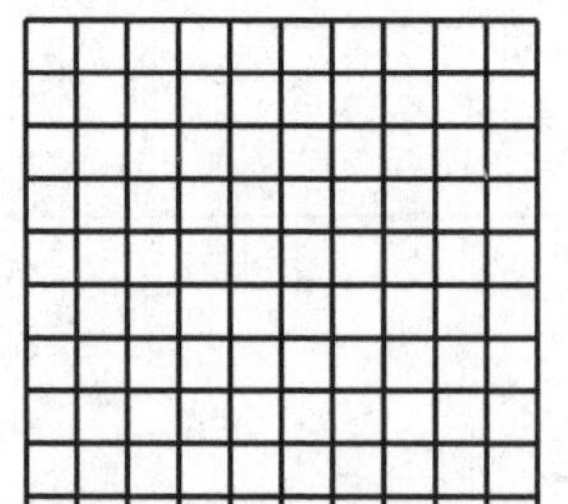

Fraction: $\frac{27}{100}$

Decimal: ______________

2. 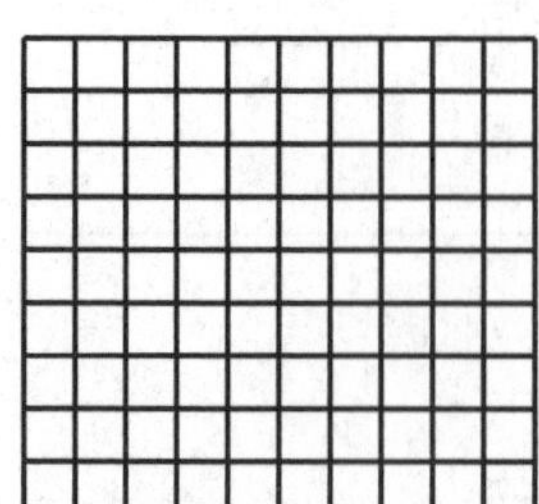

Fraction: ______________

Decimal: 0.19

3. 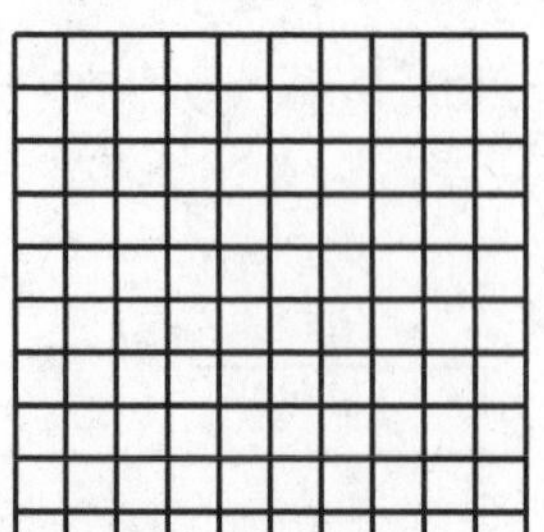

Fraction: ______________

Decimal: 0.83

4. 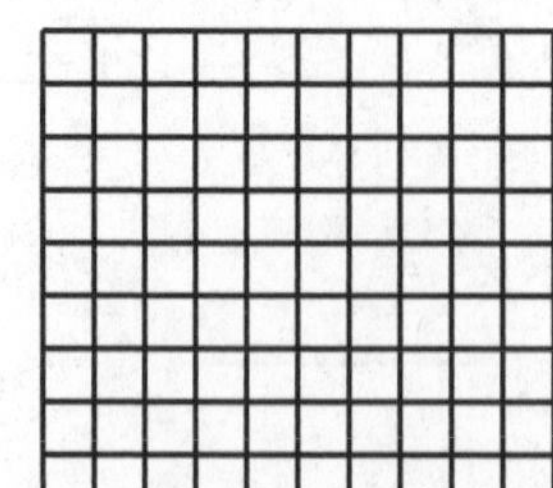

Fraction: $\frac{39}{100}$

Decimal: ______________

Name Date

Equivalent Common Fractions and Decimal Fractions

1. Shade each large square to show the fraction written next to it.

0.4

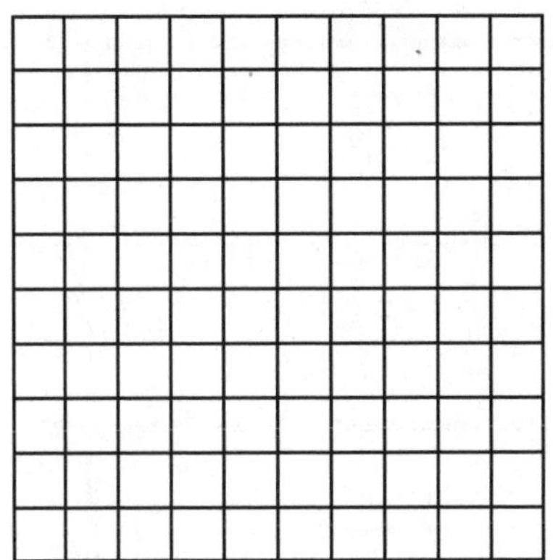

$\frac{2}{5}$

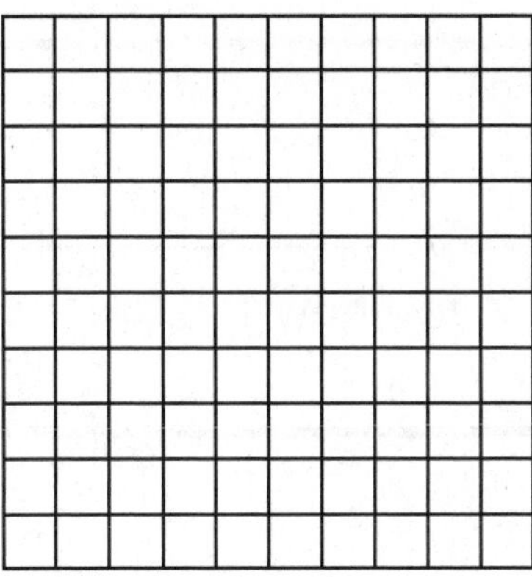

0.25

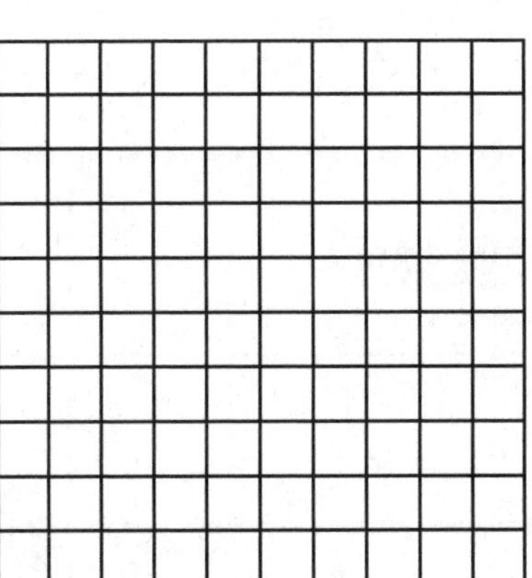

$\frac{1}{4}$

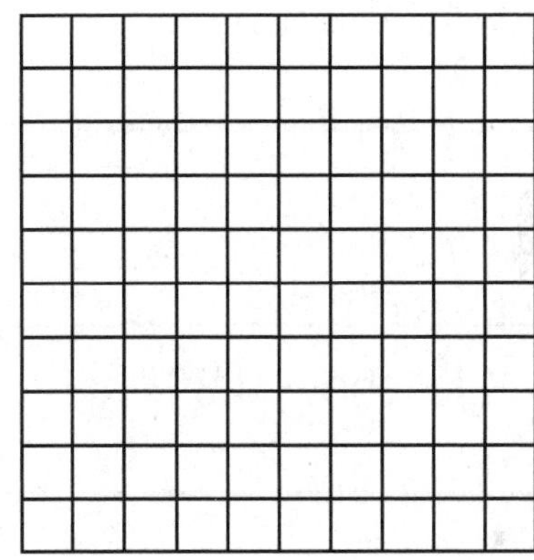

What did you notice about each pair of grids you shaded? ____________

____

2. Solve these fraction problems. You can draw pictures to help.

a. How many tenths are equal to three-fifths? _________

b. If you have a board $4\frac{1}{2}$ meters long, how many boards one-tenth of a meter long can you cut from it? ____________

Name Date

Decimals on a Number Line

Mark an ✗ to show 0.25.

Mark an ✗ to show 0.5.

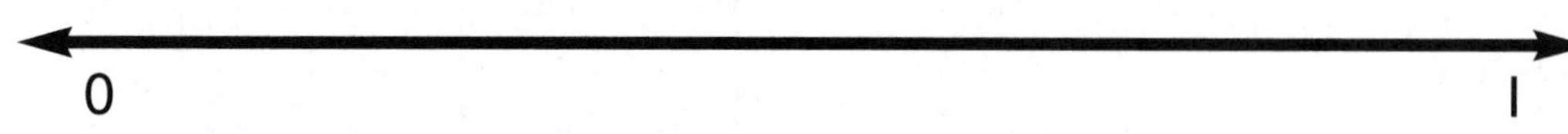

Mark an ✗ to show 0.66.

Mark an ✗ to show 0.89.

Mark an ✗ to show 1.25.

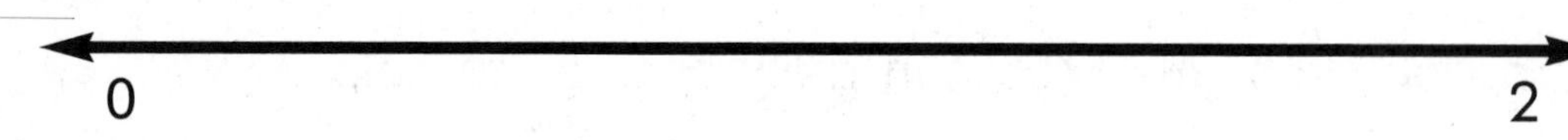

Mark an ✗ to show 3.57.

Mark an ✗ to show 5.01.

0 6

Name Date

Problems to Share

Read each story problem below.
Draw a picture to help you solve it. Write the number fact.

There are 24 tennis balls to pack in containers. A container holds 4 balls. How many containers will be needed? ________

There are 3 garden beds. 12 plants are growing in each garden bed. How many plants are there in all? ________

There are 4 children in a group. Each child has 20 marbles. How many marbles does the whole group have? ________

3 children are given $18 to share. How much spending money will each child receive? ________

A carpenter has to build 2 walkways. 14 planks are used on each walkway. How many planks will be used? ________

An egg tray holds 30 eggs. 6 eggs fit in each row. How many rows does the egg tray have? ________

Name Date

Sharing Problems

1. Use cubes to act out each problem. Write about what you discovered.

Share 23 balls equally among 4 containers. ______________________

____

Share $30 among 4 people. ______________________

____

Sort 9 planks into 2 equal piles. ______________________

____

Share 25 pencils among 4 people. ______________________

____

Share 6 large pizzas among 4 tables of people. ______________________

____

2. Fill in the missing numbers.

Amount to be shared	Number of people sharing	Amount in each share	Amount left over
48	9		3
36		7	1
	5	3	2
$42	8		
	5	6	
$27	5		

Name Date

Relating Multiplication and Division

Write the answer for the multiplication fact.
Then write two related division facts.

4 × 5 = 20	20 ÷ 5 = 4	20 ÷ 4 = 5
3 × 6 = ______	______	______
4 × 7 = ______	______	______
5 × 8 = ______	______	______
9 × 3 = ______	______	______
6 × 4 = ______	______	______
7 × 5 = ______	______	______
3 × 5 = ______	______	______
3 × 7 = ______	______	______
4 × 9 = ______	______	______
5 × 6 = ______	______	______
9 × 5 = ______	______	______

Name Date

Relating Multiplication and Division

Write the answer for the multiplication fact.
Then write two related division facts.

4 × 3 = 12	12 ÷ 3 = 4	12 ÷ 4 = 3
5 × 7 = ______	______	______
6 × 2 = ______	______	______
8 × 6 = ______	______	______
7 × 2 = ______	______	______
6 × 7 = ______	______	______
4 × 6 = ______	______	______
8 × 3 = ______	______	______
9 × 6 = ______	______	______
6 × 8 = ______	______	______
8 × 9 = ______	______	______
8 × 7 = ______	______	______

This page reviews the connection between multiplication and division. When students need to use division facts, the link to multiplication helps them to figure out any facts they forget.

Name Date

Multiplication Machines

Write the missing numbers for these multiplication "machines."

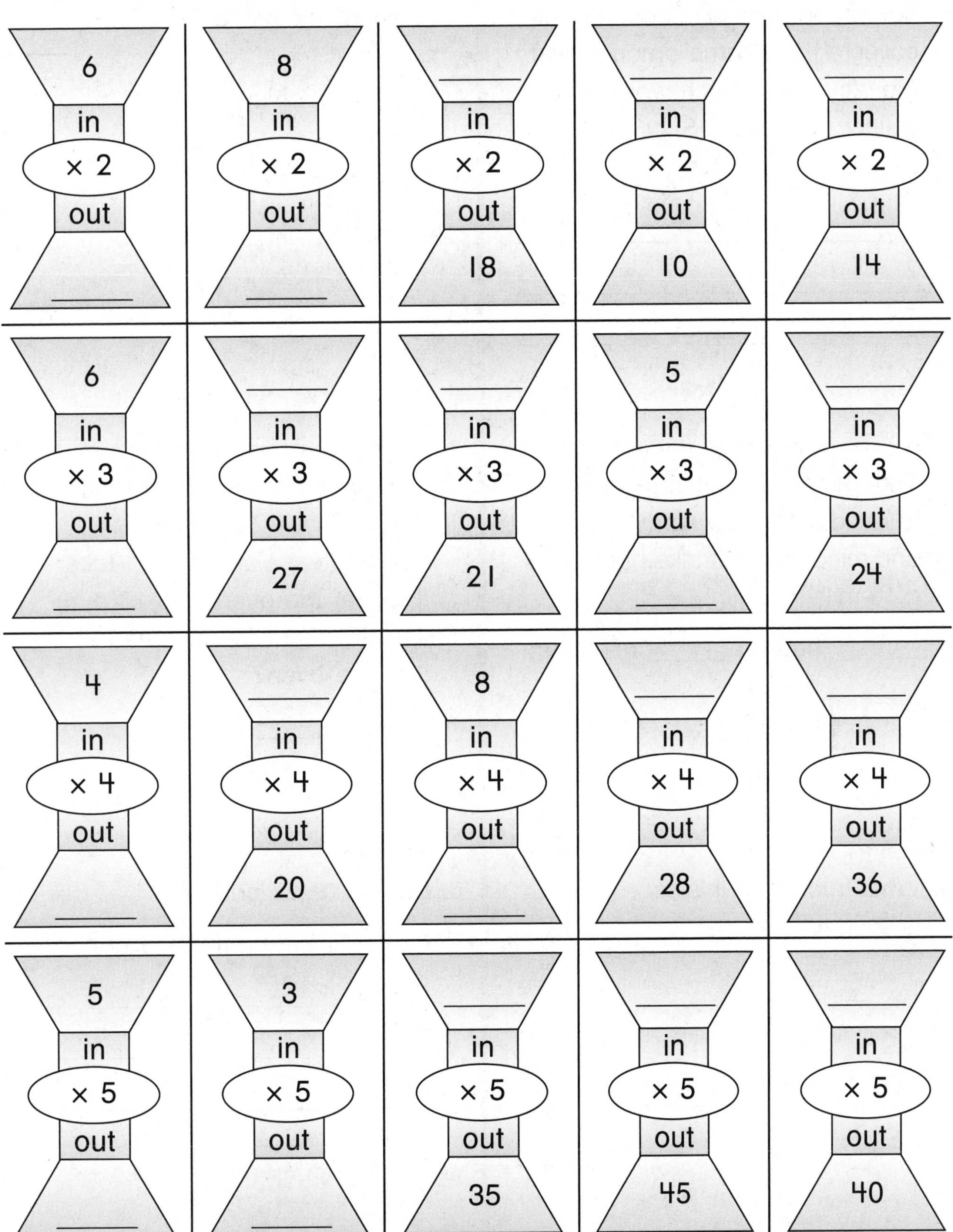

Name Date

Maintaining Concepts and Skills

1. Donyale collected $8 to deliver newspapers on Mondays. If she collected the same amount every Monday, how much would she collect in a year? $________

How did you figure it out?

____

____

____

2. Draw a shape that has exactly 2 lines of symmetry.

3. Tom paces himself at $8\frac{1}{2}$ minutes to run a mile. How many hours and minutes will it take him to run a 10-mile race?

______ hour(s) _____ minute(s)

4. The recipe says to roast the turkey for 3 hours and 20 minutes. You want to take it out of the oven at 6:15 p.m. At what time do you need to put it in the oven?

____

5. What fraction of this set is shaded?

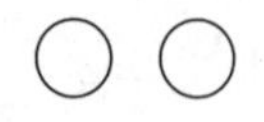

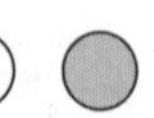
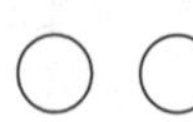

____

6. Use 3, 4, 5, and 5.

Make all the numbers you can.

____

____

____

Name Date

Multiplication Machines

Write the missing numbers for these multiplication "machines."

in	____	____	9	____	6
rule	× 6	× 6	× 6	× 6	× 6
out	30	48	____	42	____

in	____	6	____	____	____
rule	× 7	× 7	× 7	× 7	× 7
out	21	____	56	42	35

in	____	4	____	____	____
rule	× 8	× 8	× 8	× 8	× 8
out	40	____	24	56	48

in	____	____	9	____	____
rule	× 9	× 9	× 9	× 9	× 9
out	27	36	____	45	63

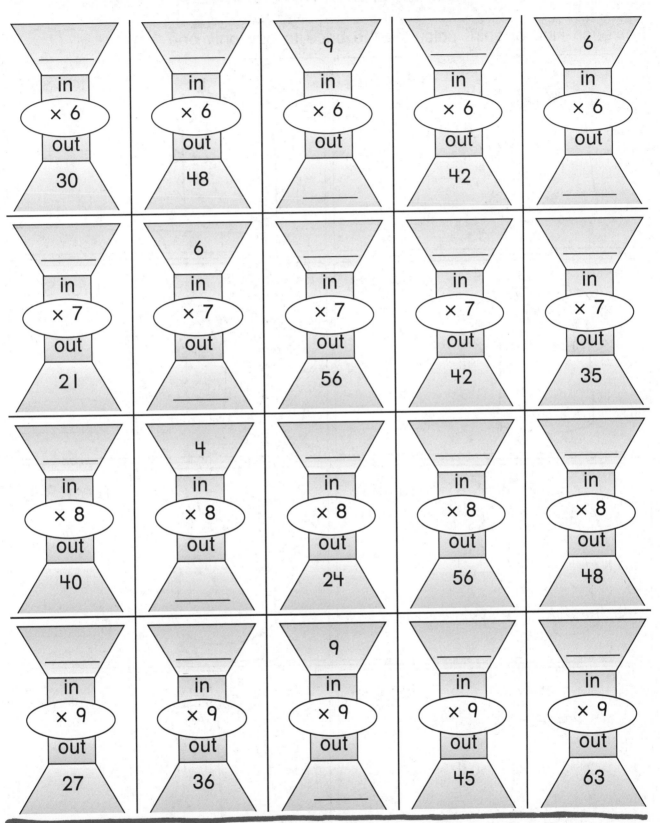

The students enjoy working with input-output "machines." This is a fun way to practice multiplication and division facts and reinforces the connection between the two operations.

Name Date

Super Shirt Factory Sale

SUPER SHIRT FACTORY SALE

For each kind of shirt, calculate the price to buy only one.

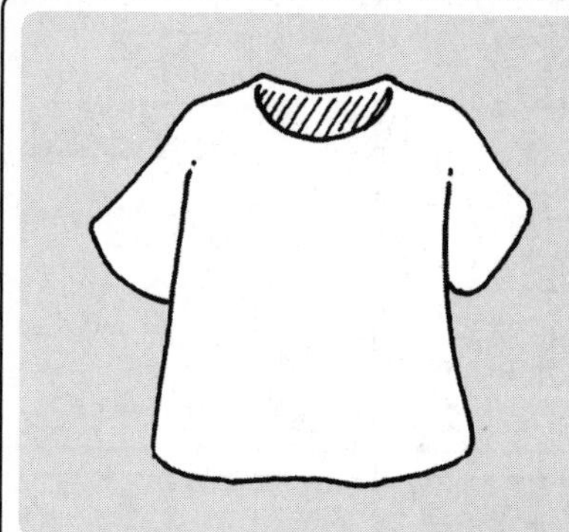

3 for $84

$\overline{)\quad\quad}$

$______

4 for $52

$\overline{)\quad\quad}$

$______

2 for $76

$______

3 for $96

$______

4 for $48

$______

8 for $96

$______

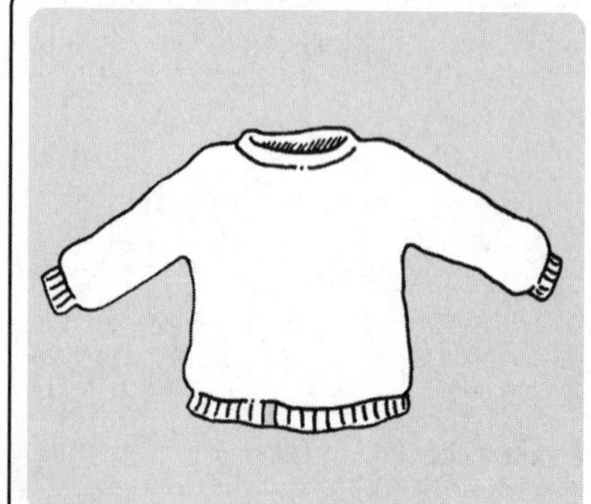

2 for $88

$______

3 for $87

$______

Name Date

Exploring Division Patterns

Divide by 2. Then divide the answer by 3.

18	12	24	30	42	36	48	54
÷ 2	÷ 2	÷ 2	÷ 2	÷ 2	÷ 2	÷ 2	÷ 2
9							
÷ 3	÷ 3	÷ 3	÷ 3	÷ 3	÷ 3	÷ 3	÷ 3
3							

Divide by 6.

	18	12	24	30	42	36	48	54
÷ 6								

Divide by 2. Then divide the answer by 4.

8	16	32	24	40	56	64	48
÷ 2	÷ 2	÷ 2	÷ 2	÷ 2	÷ 2	÷ 2	÷ 2
4							
÷ 4	÷ 4	÷ 4	÷ 4	÷ 4	÷ 4	÷ 4	÷ 4
1							

Divide by 8.

	8	16	32	24	40	56	64	48
÷ 8								

Name Date

Maintaining Concepts and Skills

1. You have 5 flavors of ice cream and 2 kinds of cones. How many different combinations can you make that have one scoop of ice cream?

__________ combinations

2. Draw the lines of symmetry in these shapes.

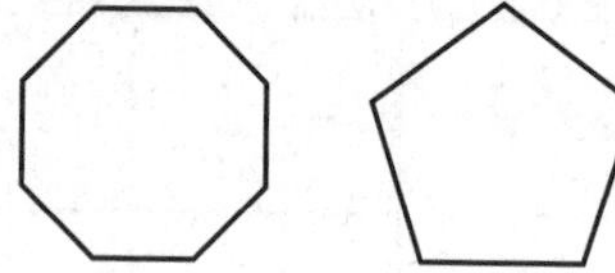

3. Solve these.

4 × 18 = _____ 6 × 19 = _____

5 × 28 = _____ 5 × 29 = _____

4. What number is 100 more than 23,346?

____

5. Use this flight schedule to answer the questions. All times are in the same time zone.

Flight No.	Departs	Arrives
348	8:05 a.m.	10:40 a.m.
451	11:18 a.m.	2:10 p.m.

How much longer is Flight 451 than Flight 348?

____

Flight 451 was 34 minutes late arriving yesterday. What time did it arrive?

____

6. Joel begins work each day at 5:15 p.m. He finishes work at 9:45 p.m. He works 5 days each week. How many hours does Joel work in 2 weeks?

__________ hours

Name Date

Grab the Cubes

Students played **Grab the Cubes** in groups of 4. Each student grabbed a handful of cubes and then recorded the number of cubes in their grab.

1. Write the answers to these questions in the spaces below each chart.
   a. How many students were in the group?
   b. What was the total number of cubes they grabbed?
   c. When they shared the cubes, how many did each student get?

Group A	Keisha	Sean	Denzel	Melissa	Toya
	6	12	9	8	10

Number of students ________ Total number of cubes ________

Number of cubes in each share ________

Group B	Tiffani	Engracia	Rashawn	Amanda
	7	9	13	7

Number of students ________ Total number of cubes ________

Number of cubes in each share ________

Group C	Calvin	Ann	Jamila	Luz	Daniel	Beto
	9	6	8	7	8	10

Number of students ________ Total number of cubes ________

Number of cubes in each share ________

2. Play your own **Grab the Cubes** game. Record your results in the chart below.

________	________	________	________	________

Number of students ________ Total number of cubes ________

Number of cubes in each share ________

Name

Date

Maintaining Concepts and Skills

1. Draw the lines of symmetry in these shapes.

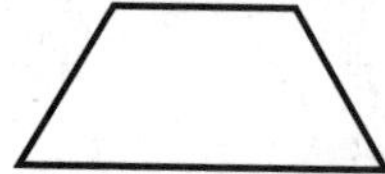

2. **A** is a line of symmetry. Draw the right side of the picture so that it balances.

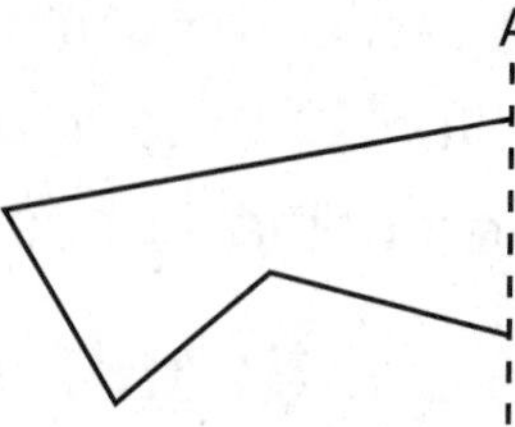

3. Nate left for work at 7:50 a.m. He did not get home until 4:25 p.m. How long was he away that day?

_________ hours _________ minutes

4. How can you use the answer to 20×42 to help find the answer to 22×42?

____

____

____

5. Each day Delsonia works from 8:15 a.m. until 11:00 a.m. and from 11:45 a.m. until 3:00 p.m. How many hours does she work in a 5-day week?

_____________ hours

6. Use the information in question 5. If Delsonia gets paid $5 per hour, how much does she earn each week?

$_____________

Name Date

Exploring Division Patterns

Divide by 2. Then divide the answer by 2.

12	8	16	24	20	36	32	28
÷ 2	÷ 2	÷ 2	÷ 2	÷ 2	÷ 2	÷ 2	÷ 2
÷ 2	÷ 2	÷ 2	÷ 2	÷ 2	÷ 2	÷ 2	÷ 2

Divide by 4.

	12	8	16	24	20	36	32	28
÷ 4								

Divide by 2. Then divide the answer by 5.

20	30	70	40	60	80	50	90
÷ 2	÷ 2	÷ 2	÷ 2	÷ 2	÷ 2	÷ 2	÷ 2
÷ 5	÷ 5	÷ 5	÷ 5	÷ 5	÷ 5	÷ 5	÷ 5

Divide by 10.

	20	30	70	40	60	80	50	90
÷ 10								

When your child has completed the page, ask what he or she notices about the answers. (For example, dividing by 10 is the same as dividing by 2 and then by 5.) You could then ask questions such as, "What is an easy way of dividing by 15?" (Divide by 3 and then by 5.)

Name

Date

Maintaining Concepts and Skills

1. Your dad has agreed to take you and 4 friends to a concert. It starts at 8:00 p.m. and it takes 1 hour and 25 minutes to get there. You need 10 minutes to pick up each friend. At what time should you start?

____

2. Solve these.

8 × 25 = _____

4 × 112 = _____

7 × 59 = _____

39 × 6 = _____

3. Some astronauts went into space on April 21 and returned to Earth on November 11. How many days were they in space?

_______ days

4. Sonya's parents drove 55 miles per hour for 8 hours before they got to the city. About how far was the trip?

_______ miles

5. 5 × 25 = _____ 5 × 35 = _____

10 × 25 = _____ 10 × 35 = _____

20 × 25 = _____ 20 × 35 = _____

What patterns do you see in the solutions you found?

____

____

____

6. Draw a shape that has 2 or more lines of symmetry.

How do you know?

____

____

____

Name Date

Solving Division Problems

Divide. Write the answers.

$2 \overline{)14}$ = 7	$5 \overline{)15}$	$4 \overline{)24}$	$3 \overline{)18}$
$2 \overline{)14 \text{ tens}}$ = 7 *tens*	$5 \overline{)15 \text{ tens}}$	$4 \overline{)24 \text{ tens}}$	$3 \overline{)18 \text{ tens}}$
$2 \overline{)140}$ = 70	$5 \overline{)150}$	$4 \overline{)240}$	$3 \overline{)180}$
$4 \overline{)12}$	$5 \overline{)35}$	$3 \overline{)24}$	$6 \overline{)24}$
$4 \overline{)12 \text{ tens}}$	$5 \overline{)35 \text{ tens}}$	$3 \overline{)24 \text{ tens}}$	$6 \overline{)24 \text{ tens}}$
$4 \overline{)120}$	$5 \overline{)350}$	$3 \overline{)240}$	$6 \overline{)240}$
$4 \overline{)36}$	$6 \overline{)54}$	$5 \overline{)30}$	$6 \overline{)42}$
$4 \overline{)36 \text{ tens}}$	$6 \overline{)54 \text{ tens}}$	$5 \overline{)30 \text{ tens}}$	$6 \overline{)42 \text{ tens}}$
$4 \overline{)360}$	$6 \overline{)540}$	$5 \overline{)300}$	$6 \overline{)420}$

Name Date

Solving Division Problems

Divide. Write the answers.

2) 16	3) 15	4) 16	2) 18
2) 16 tens	3) 15 tens	4) 16 tens	2) 18 tens
2) 160	3) 150	4) 160	2) 180
3) 27	4) 32	5) 45	6) 36
3) 27 tens	4) 32 tens	5) 45 tens	6) 36 tens
3) 270	4) 320	5) 450	6) 360
6) 48	5) 40	4) 28	3) 21
6) 48 tens	5) 40 tens	4) 28 tens	3) 21 tens
6) 480	5) 400	4) 280	3) 210

In class, students are beginning to explore the division algorithm. Here they are using patterns to help them recognize the "tens" in 3-digit numbers. Soon they will divide numbers that are not multiples of 10.

Name Date

Reading Numbers on a Number Line

For each arrow, write the number the arrow is pointing to.

5,020

5,010

5,000

30,100

30,000

29,900

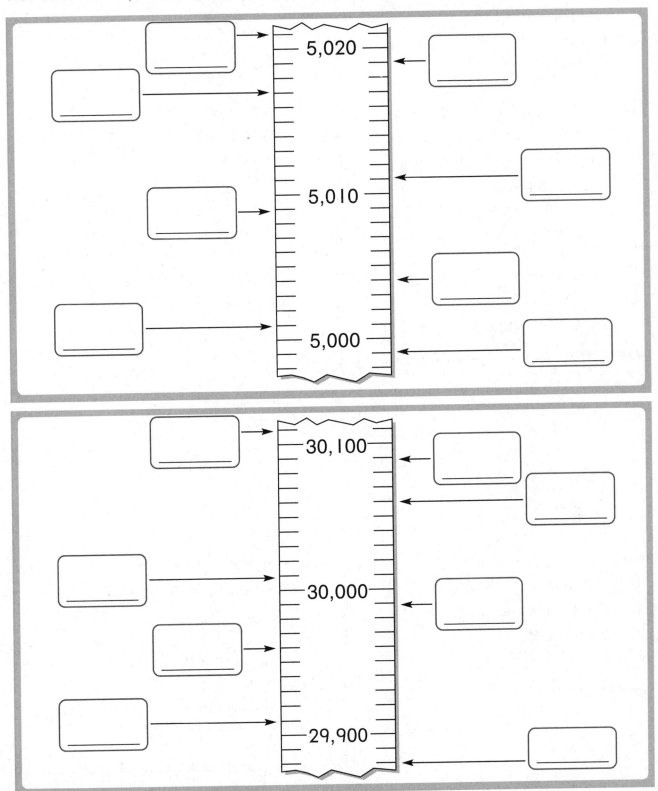

The number lines on this page provide practice in sequencing numbers and counting backward by ones, tens, and hundreds. Before your child writes the answers, ask him or her to point to each mark on the number line and say the number it represents.

Name Date

Maintaining Concepts and Skills

1. a. 7 bags of marbles with 8 marbles in each. How many marbles?

_____ marbles

b. Senta's uncle gave her 3 quarters, 4 dimes, and 3 pennies. What is the total value of that money?

_____

2. a. Write the answers.

10 × 6 = _____ 11 × 6 = _____

12 × 6 = _____ 13 × 6 = _____

b. Juan found 12 boxes of pencils. Each box had a dozen pencils. How many pencils did he find?

_____ pencils

3. Show 3 more ways to change a $5 bill into dollars, quarters, and dimes. Use at least one of each.

dollars	quarters	dimes
3	6	5

4. Draw all the balance lines in the shapes.

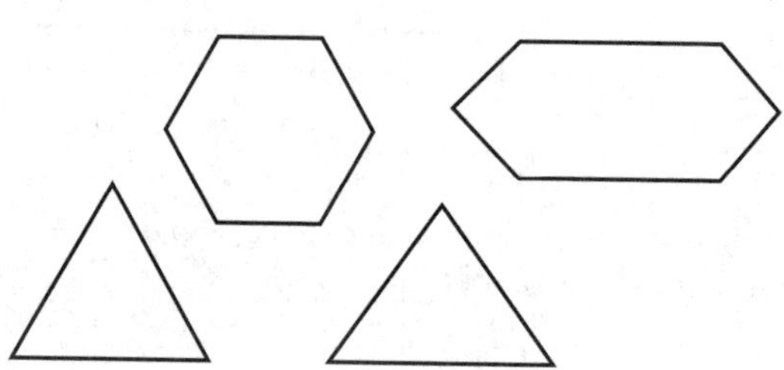

5. 5 children get fair shares of 6 apples. How many apples does each get?

_____ apples

6. Estimate or solve mentally.

13 × _____ = 143

15 × _____ = 210

17 × _____ = 102

Explain your thinking.

_____

_____

Name Date

Finding the New Amount

1. For each price tag, write the prices that are $1 more and $1 less.

$1 more	$______	______	______	______
Starting price	$3,289	$64,500	$71,569	$85,999
$1 less	$______	______	______	______

2. For each price tag, write the prices that are $100 more and $100 less.

$100 more	$______	______	______	______
Starting price	$70,329	$50,061	$17,000	$50,940
$100 less	$______	______	______	______

3. For each price tag, write the prices that are $1,000 more and $1,000 less.

$1,000 more	$______	______	______	______
Starting price	$49,730	$52,010	$30,009	$79,302
$1,000 less	$______	______	______	______

4. Write how each number was changed to make the new display below it.

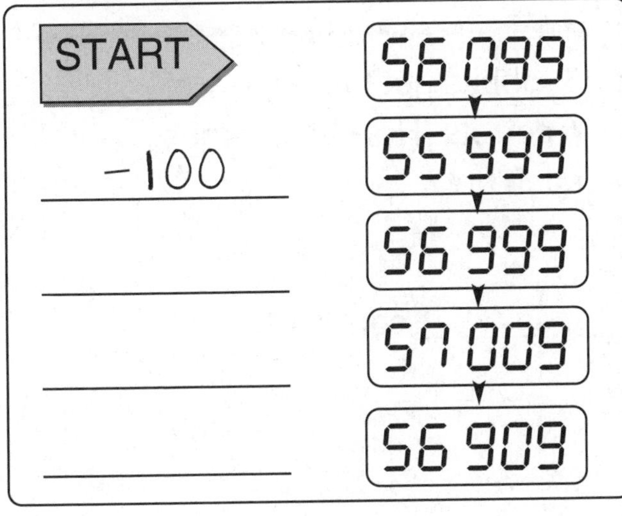

START

79 909

______

80 909

______

81 009

______

80 999

______

80 899

Name Date

Maintaining Concepts and Skills

1. Ring the problems that have an answer greater than 150. Estimate or solve mentally.

$$\begin{array}{r} 48 \\ \times 3 \\ \hline \end{array} \quad \begin{array}{r} 78 \\ \times 2 \\ \hline \end{array} \quad \begin{array}{r} 39 \\ \times 4 \\ \hline \end{array} \quad \begin{array}{r} 52 \\ \times 2 \\ \hline \end{array}$$

10×14 6×26 8×18 11×14

2. Look at the price tags. Ring the prices of 3 things you can buy if you spend close to $10.00.

$1.40 $2.65 $0.98

$6.55 $3.10 $1.97

3. Marla collected dimes for the food fund. She collected 15 dimes in the first block, 26 dimes in the second block, and 39 dimes in the third block. How many dimes did she collect?

__________ dimes

4. Estimate the total amount of clay these children used. Adam used $\frac{1}{3}$ bar. Elana used $\frac{1}{4}$ bar. Thian used $\frac{1}{2}$ bar. Saul used $\frac{3}{4}$ bar.

__________ bars of clay

5. Pedro found 24 shells for his collection. He put 8 in each box. How many boxes did he use?

__________ boxes

6. 8 rabbits get fair shares of 6 carrots. How much does each get?

__________ carrot

Name Date

Calculating Sums and Differences

Write the number you would see on the calculator after these keys are pressed.

Starting number

27809

Keys		Result
+ 1 0 0 0 0	=	37,809
+ 1 0 0 0	=	38,809
+ 1 0 0	=	
− 1 0 0 0 0	=	
+ 1 0 0	=	
+ 1 0 0 0	=	
+ 1	=	
− 1 0 0	=	
+ 1 0 0 0 0	=	
+ 1 0 0	=	
− 1 0	=	
− 1 0 0 0	=	

Name Date

HOME LINK

Calculating Sums and Differences

Write the number you would see on the calculator after these keys are pressed.

Starting number: 42090

−	1 0 0 0 0	=	32,090
−	1 0 0 0	=	31,090
−	1 0 0 0	=	
+	1 0 0	=	
−	1 0 0 0	=	
−	1	=	
+	1 0	=	
+	1 0 0 0 0	=	
+	1 0 0 0	=	
−	1 0	=	
−	1 0 0 0	=	
−	1 0 0	=	

The activities on this page help students focus on place value for 5-digit numbers and provide practice in mental addition and subtraction. When your child has finished, he or she can use a calculator to check the answers.

Name Date

Comparing Populations in Ohio

This chart shows the populations of some cities in Ohio.

City	Population	City	Population
Bowling Green	28,176	Middletown	48,590
Canton	80,806	Newark	48,245
Chillicothe	21,796	Sandusky	28,223
Findlay	38,967	Springfield	70,487
Lima	42,382	Warren	50,793

1. Which city has the **greatest** population? ______
2. Which city has the **least** population? ______
3. Which city has a population **closest** to 75,000? ______
4. Which cities have a population **less than** 30,000?

   ______

5. Which cities have a population **greater than** 50,000?

   ______

6. Which city has a population **about**

   half the size of Newark? ______

   twice the size of Findlay? ______

   20,000 more than Warren? ______

7. Rewrite the chart, placing the cities in order from **greatest** to **least**.

1. ______	6. ______
2. ______	7. ______
3. ______	8. ______
4. ______	9. ______
5. ______	10. ______

8. Mansfield was left off the chart. It has a population of 51,600.
   What position would it have in your list? ______

Name

Date

Maintaining Concepts and Skills

1. Show 3 other ways you can use coins and bills to make $30.50.

$10 bills	$5 bills	quarters	dimes	nickels
1	3	20	3	4

2. a. Ernest bought 3 oranges, 2 apples, and 5 bananas. How much did he spend?

$__________

b. Maria had $2.50. What fruit could she buy?

____

____

3. Mr. Baker had 40 donuts. He gave one-fifth to Tan. He gave 10 to Sella. He gave 12 to Alice, Dela, and Sue to share.

a. What fraction of the 40 donuts did Mr. Baker give away?______

b. Who got the most donuts?

____

4. 5 of the 23 students in the class are wearing sneakers. About what fraction of the students is that?

______

How did you decide?

____

____

5. Jeff had 3 coins with a value of 16 cents. What were the coins?

____

6. Xia helped the teacher arrange the chairs. She put 5 chairs at each of the 6 tables. How many chairs did she use?

______ chairs

Name Date

Using Data

Flyer I	747
First flight: December 17, 1903	**First flight:** February 9, 1969
Wingspan: 40 ft 4 in.	**Wingspan:** 195 ft 8 in.
Length: 21 ft 1 in.	**Length:** 231 ft 4 in.
Take-off weight: 750 lb	**Take-off weight:** 710,000 lb
Altitude: 12 ft	**Altitude:** 15,500 ft
Flight duration: 12 secs	**Flight duration:** 76 mins

1. How many years ago was the first flight of
   a. Flyer I? ______________
   b. the 747? ______________
2. How many years apart were the 2 flights? ______ years
3. Look at the **wingspan** and **length** of Flyer I.
   a. Which dimension is greater? ______________
   b. What is the difference between the 2 dimensions? ______________
4. Look at the **wingspan** and **length** of the 747.
   a. Which dimension is greater? ______________
   b. What is the difference? ______________
5. Find this information about both flights. Write about the difference.
   a. Take-off weight ______________
   b. Altitude ______________
   c. Flight time ______________
6. On its first flight, Flyer I flew 170 ft. Compare this with the length of the 747. What did you discover? ______________

____

Name Date

Maintaining Concepts and Skills

1. Write a story problem for each.

3 × 12 = ______

______

______

______

______

10 × 21 = ______

______

______

______

______

2. a. What could you buy for about $45.00? (You can buy more than one of the same item.)

______

______

b. If you bought 3 different items, what is the least you could spend? $______

The most? $______

3. How much change?

Cost	Amount given	Change
$1.23	$5.00	
$3.67	$20.00	
$9.47	$20.00	

4. Four friends shared 2 dozen cookies. How many cookies did each get? ______ cookies

What fraction of the cookies did each person get?

______

5. 3 children shared the cost. About what was a fair share?

______ ¢

6. Draw a triangle that is **3** cm on each side. Draw in all its lines of symmetry.

Name Date

Ordering 6-digit Numbers

Write each set of numbers in order from **least** to **greatest**.

3,476	4,736	3,746	4,367
______	______	______	______
12,570	12,057	12,507	12,075
______	______	______	______
20,109	19,020	19,002	20,019
______	______	______	______
200,199	199,020	199,200	199,002
______	______	______	______
191,091	191,910	191,190	191,901
______	______	______	______
991,910	991,901	991,109	991,091
______	______	______	______

Name Date

Ordering 6-digit Numbers

Write each set of numbers in order from **least** to **greatest**.

5,109	5,190	5,091	5,019
______	______	______	______
17,910	17,901	17,091	17,190
______	______	______	______
30,100	30,010	31,000	30,001
______	______	______	______
121,012	120,211	121,120	120,112
______	______	______	______
500,194	500,149	491,500	491,050
______	______	______	______
901,091	901,109	910,019	901,190
______	______	______	______

Ordering numbers is an important mathematical skill. After he or she has completed the page, ask your child to tell how he or she decided which number was the least in each set.

Name Date

Working with One Million

1. Calculate how much money you would have if you had one million

2. Write 2 things you know about one million.

____

____

3. Write where you might see one million things in your

school ______________________

city park ______________________

town ______________________

4. How many of these would you need to make one million?

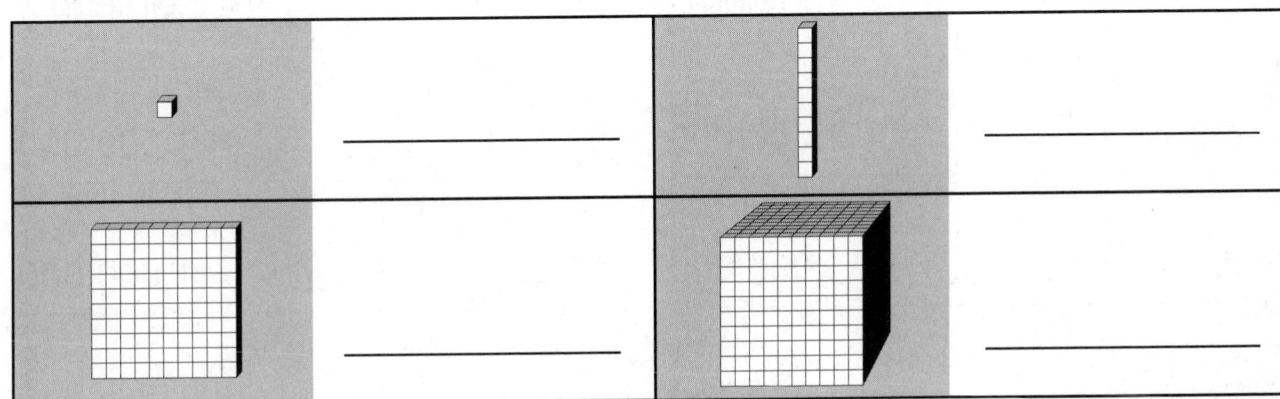

5. Calculate how old you would be if you lived for

one million hours. ______________________

one million minutes. ______________________

one million seconds. ______________________

Name Date

Adding on to Make One Million

Write the missing numbers.

3,000 + ________ = 10,000	3,750 + ________ = 10,000
3,000 + ________ = 100,000	3,750 + ________ = 100,000
3,000 + ________ = 1,000,000	3,750 + ________ = 1,000,000
78,000 + ________ = 100,000	27,000 + ________ = 100,000
78,000 + ________ = 1,000,000	27,000 + ________ = 1,000,000
91,500 + ________ = 100,000	77,700 + ________ = 100,000
91,500 + ________ = 1,000,000	77,700 + ________ = 1,000,000
49,500 + ________ = 100,000	51,500 + ________ = 100,000
49,500 + ________ = 1,000,000	51,500 + ________ = 1,000,000
925,000 + ________ = 1,000,000	850,000 + ________ = 1,000,000
775,000 + ________ = 1,000,000	550,000 + ________ = 1,000,000
865,000 + ________ = 1,000,000	325,000 + ________ = 1,000,000

Encourage your child to work mentally to figure out the missing numbers. The patterns on the page help to reinforce place-value concepts for thousands and millions.